Remote Sensing of the Mine Environment

Remote Sensing of the Mine Environment

Editor

Sanjay Singh

scitus
academics

Remote Sensing of the Mine Environment

Edited by **Sanjay Singh**

Printed in 2017

ISBN: 978-1-68117-493-8

Library of Congress Control Number: 2015936600

SCITUS Academics LLC,
616, Corporate Way, Suite 2, 4766,
Valley Cottage, NY 10989

www.scitusacademics.com

Notice

Contents

Preface

This book discussed on the basic principles of remote sensing and its applications in mine environment monitoring. Building from a reader's basic knowledge of mine monitoring, it teaches how to implement remote sensing techniques and how to interpret the acquired data for different purposes. Following a general introduction to remote sensing principles and image analysis, mine subsidence monitoring, slope stability monitoring, reclamation planning and implementation, and post-closure mine and land use analysis are explained and illustrated. With the help of case studies, the techniques and tools presented are demonstrated.

Editor

Chapter 1

Land Cover Fragmentation Using Multi-Temporal Remote Sensing on Major Mine Sites in Southern Katanga (Democratic Republic of Congo)

Laëtitia Dupin[1, 2], Collin Nkono[1], Christian Burlet[1], François Muhashi[1], and Yves Vanbrabant[1]

[1]Royal Institute of Natural Sciences, Bruxelles, Belgium
[2]Brussels Environment, Bruxelles, Belgium
[3]RBNS/Convention on Biological Diversity, Bruxelles, Belgium

ABSTRACT

The study areas are located in the Katanga province to the South Eastern part of the Democratic Republic of Congo (DRC). It focuses on the Kolwezi and Tenke-Fungurume mining centers, located in the vicinity of the Basse-Kando reserve. The 3 study areas have faced large scale human induced the fragmentation of land cover. A combination of ancillary data and satellite imageries was interpreted to construct fragmentation dynamics over the last 30 years. This study is an initial step towards assessing the impact of fragmentation on sustainable land cover in the Katanga. The results bring out that large trends of fragmentation differently occurred over the last 30 years (1979 to 2011) in the three focused areas. The most dominant fragmentation processes were gains in barren soil and cities surface and a sharp reduction in burned areas. In Kolwezi, a close relationship is observed between growth and regression of barren soil and cities over vegetation. The Tenke-Fungurume site shows a growth during the 1980-1990 time slice and regression of vegetation during the following two decades. The Basse-Kando site analyze brings out growth of vegetation and regression of burned area due to vegetation conservation efforts. This is one of the studies in Katanga around mines activities that combine multi-source and spatio-temporal data on land cover to enable long-term quantification of land cover fragmentation.

INTRODUCTION

Land cover fragmentation is now recognized as an indicator and a prominent interface between human activities and global environmental change. Human and environment interactions involve conversion of natural landscapes into anthropogenic use or by changing management practices in human-dominated landscapes [1].

Understanding the role of land use in global environmental change and the long-term human-environment interaction require historical reconstruction of past land cover conversions. The information on land cover under developed countries is usually missing, out dated or inconsistent [2,3]). The launch of the Landsat imagery platform in 1972, followed by others such as SPOT and ASTER, has provided satellite remote sensing (RS) capacity to detect LUCC over the past 30 years at most. As suggested by [4], "Land cover" refers to the biophysical state of the earth's surface and immediate subsurface [5], and, "Land fragmentation" brings out the human employment of land [6]. The combined application of land cover and fragmentation data allows detection of type of change, as well as the way land is changing [7]. In this study we will provide fragmentation quantification in land cover. Rare environmental studies have been undertaken in this area.

In this paper, the study of the environmental changes in South-Eastern of Democratic Republic of Congo (DRC) aims to establish a detailed assessment of fragmentation patterns over time and space. The study was undertaken more specifically in the surrounding of Kolwezi and Tenke-Fungurume copper-cobalt mining sites and the Basse-Kando Natural Reserve. For our study, we used the Landsat MSS, TM, ETM and SPOT images, Shuttle Radar Topography Mission (SRTM) Digital Elevation Model (DEM) and the Aster DEM. The goal of this study was to detect land cover fragmentation (magnitude, spatial and temporal analyze) in terms of their spatio-temporal pattern over a period from pre- 1980 to 2011 for the Katanga province (Figure 1(a)).

The remaining sections of this paper are organized as follow: Section 2 introduces the background of the study area; Section 3 describes the data and method used in this article; in Section 4, the land cover fragmentation results are presented, followed by a discussion of the results in Section 5; the conclusion of this research is given in Section 6.

STUDY AREA

The study areas are located in the South-Eastern part of the DRC, in the Katanga province. It focuses on the Kolwezi and Tenke-Fungurume mining centers, and the Basse-Kando reserve. The Katanga province is one of the largest and most populated provinces of DRC. It encompasses apart from Kinshasa, the capital, the major cities of the country (Lubumbashi, Kolwezi, Likasi), not only from a demographic point of view but also economically (Figure 2). The Katanga province shares the DRC border with three countries: Angola to the Southwest, Zambia to the South and Southeast and to the East Tanzania. Vast reserves of copper (Cu) and cobalt (Co) are found to the South of the Katanga province, between Kolwezi and Lubumbashi strongly encountered linked to the Katanga super-group.

Geological setting of the Katanga Super-group is commonly subdivided into three lithostratigraphic groups: the basal Roan, the middle Nguba (formerly Lower Kundelungu) and the Kundelungu (formerly Upper Kundelungu) at the top [8-14].

In the vicinity, several major reserves endure pressure. The Basse-Kando reserve is part of the conservation areas of the South-Eastern DRC. Major parks lie in the vicinity of the mining activities. The national park is under State ownership and management for the protection of flora and fauna. The reserve is a legalized land and must be managed for the benefit of all members of the community. Any resource extraction must be subjected to a management plans approved by the Government. The buffer zone is the land immediately adjacent to the boundaries and surrounds the entire protected areas (park and reserve). It is co-managed the best suitable way for and by the various stakeholders including government institutions, non-governmental organizations, communities associations and to some extent with the involvement of the private sector. This is the case of the Basse-Kando, one of the focus areas of the present study. The last decades have seen depletion of the biodiversity in the reserve as well as in its surroundings. Moreover, no environmental assessment is available in the Katanga province, which reveals a real lack in environmental management [15].

METHODOLOGY

Remote Sensing Data

Medium resolution images were used as Landsat MSS for 1979, TM for 1990 and ETM for 2000 and Spot 4

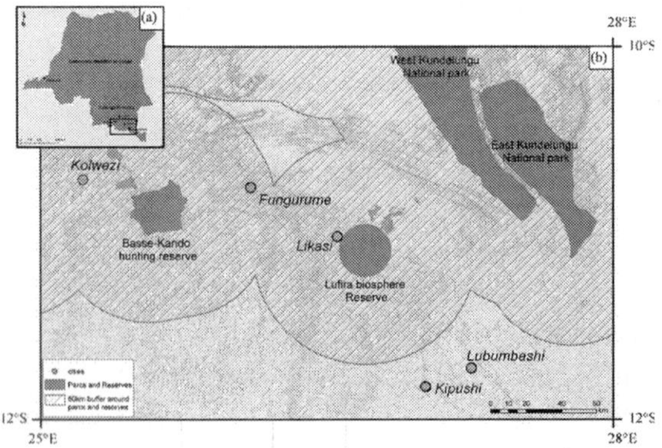

Figure 1: (a) Overview of the Katanga Province and major cities located in Democratic Republic of Congo; (b) Location of the study area, Southern Katanga parks, reserves and buffer zones.

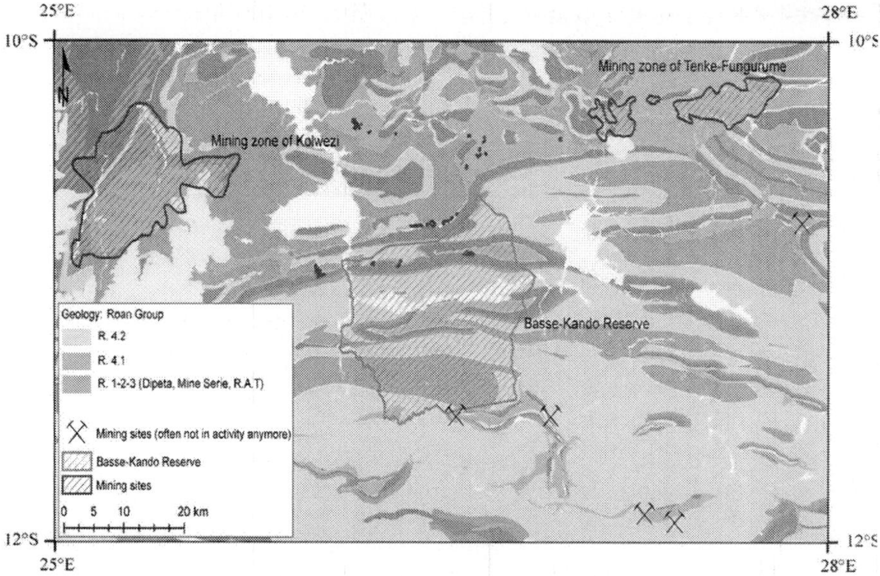

Figure 2: General geological map of Katanga indicating the main units. The Roan group (R) is the richest cobalt-cobalt mineralisation ore deposits: R1 is the lower unit, and R4 is the upper unit [14].

vegetation for 2011, with 60, 30, 30 and 20 meter resolution, respectively. The multi-temporal approach used in this study required 3 imagery scenes, one of each focused area, for 1979, 1990, 2000 and 2011. The type of image coverage was chosen considering their availability for the area, for the period of time, their similarity in term of spectral range and spatial resolution. The images were selected for the period between end of May and early June. This corresponds to the beginning of the dry season and ensures a minimum cloud cover on the images and minimizes the presence of fire patches widely spread in a later stage of the dry season.

Appropriate remote sensing geometric and radiometric corrections were applied through digital image processing techniques to investigate more accurately land cover patterns. Landsat images were acquired from University of Maryland GLCF web site [16]. They were already georeferenced but showed some spatial shifts between images for the MSS and TM. The SPOT images were acquired without any spatial reference system. The Landsat MSS and TM were re-projected and the SPOT images were georeferenced all based on the Landsat ETM+, with the WGS84 reference system. This is an important step to ensure the best spatial match for the multi-temporal land cover fragmentation analyze.

Pre-Processing of Satellite Data

Some atmospheric corrections were applied following the COST technique [17] in order to build mosaics and compare the different period images on the same spectral base. Some Landsat MSS images revealed some spurious lines. A frequently used technique to correct line-drops consists in replacing the defective pixels with average DNs (digital numbers) of adjacent pixels [18], which is efficient in the case of one or two pixel wide lines. Spectral filters were implemented based on Fourier analysis to remove most of the periodic noise [19].

The negative effect of shadowing is the visual quality degradation of the images and cause several nuisances during the analysis of these images by modifying spectral response of areas under shadow. Thus, they are often confused with other types of objects which have a similar spectral signature. The results are a misclassification of areas covered by shadow. In the study area, the shadow spectral signature is confused with water and burning signatures. It is important to correct this effect to improve the classification process. In this purpose, the c-correction method was applied to the images using the SRTM and the Aster DEM.

Image Classification

Here we used traditional Maximum Likelihood Classifier (MLC) because of its ease in simple operation, application and good performance (Figure 3). We propose a standard classification strategy that uses different input feature sets, unlike conventional classification methods using identical classifier structure and feature sets to each land cover type.

Land Cover Type, Change and Fragmentation Analysis

The training sites were chosen to be classified as follow:

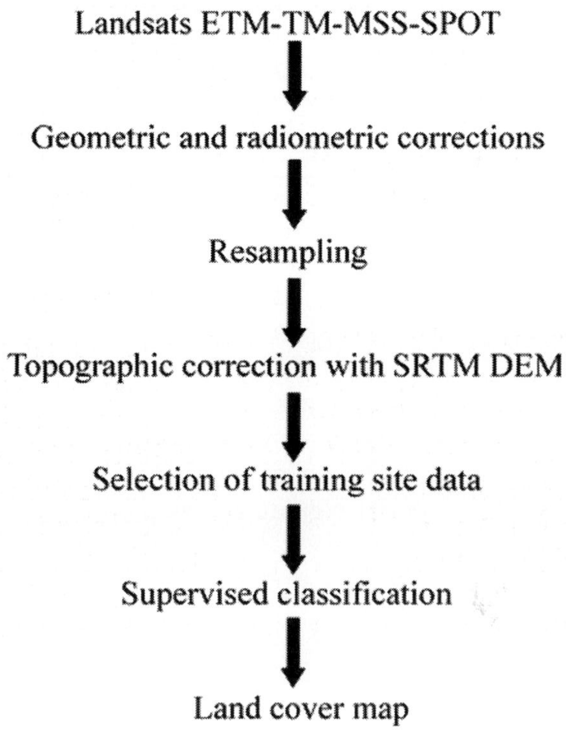

Figure 3: Methodological flowchart.

1) vegetation, 2) barren soil and cities, 3) burned area and 4) river. Then changes between the various time slices were analyzed and quantified.

Maps were analyzed using ArcView (version 9; ESRI (1999)) and its extension Arc View Spatial Analyst for Windows to quantify land cover changes, vegetation loss and to configure grid covers for the application of landscape spatial indices. These indices (or spatial metrics) were computed by FRAGSTATS (version 4.1) [20,21] to compare the spatial pattern of vegetation, barren soil and cities, and burned area cover for each time interval and study area.

The spatial configuration of land cover fragments was quantified and compared on the basis of the following key landscape metrics selected after reviewing recent vegetation fragmentation studies [22-29]: 1) patch area (%), 2) largest patch index (% of the landscape comprised by the largest patch), 3) aggregation index (% likely hood adjacencies between cells of the same patch type) and 4) number of patches. All of these metrics have been applied in the assessment of spatial attributes in fragmented landscapes.

Accuracy Assessment

According to the rules suggested by PCI geomatics for ranges of separability values of class signatures, a "good" signature separability was achieved for each image classified, the average Bhattacharrya distance ranging from 1.979 to 1.987. Overall agreement of classification was 83.2% for the 1979 MSS, 84.3% for 1990 TMimage, 85.9% for the 2000 ETM image, and 84.9% for the 2011 SPOT image. The lowest values of accuracy corresponded to shrubland, arboreus shrubland, and native vegetation categories. These three groups are very similar in their spectral signatures because they correspond to stages in a continuous succession process, which may easily produce misclassifications between the categories assigned in the training site and those classified by the algorithm, which leaded us to group them all in one category (vegetation).

RESULTS

Changes in the landscape distribution in the three study areas for the 4 time slices were quantified as difference between image pixels of 2 successive periods (geometric and radiometric normalization). The results of the land use/land cover classification for the time series are presented in Table 1.

Change in Land Cover

Land cover fragmentation were analyzed using the area statistics (Table 1) derived from land cover type maps (Figure 4). In 1980, the vegetation covered 40% of the territory in Kolwezi (Figure 5(a)), about

70% of the territory in Tenke-Fungurume (Figure 5(b)), and 50% of the land on the Basse-Kando site (Figure 5(c)). The next decade (1990) is marked by increased the area covered by vegetation on the three study sites, it reaches 50% in Kolwezi ((Figure 5(a)), 85% in Tenke-Fungurume (Figure 5(b)), and this increase is rapid on the site of the Basse-Kando with 80% (Figure 5(c)). During the second decade, the dynamics of change is not the same on all three sites. The surface covered by vegetation undercome a regression phase in Kolwezi from 50% to 40%, the regression is less important on Tenke-Fugurume where it then covers a little less than 80%, unlike the BasseKando that surface covered by vegetation exceeds 80%. The third decade is marked by a slight regeneration of the vegetation in Kolwezi, while the regression continues in Tenke-Fungurume, and the Basse-Kando site known the first phase phenomena of regression.

The second type of coverage is the barren soil and cities; its dynamics of occupancy is shown in the Figures 5(a)-(c). Since 1980, the barren soil and cities occupies 40% of the land surface in Kolwezi, less than 10% at Tenke-Fungurume, and the Basse-Kando reserve. The next decade sees the territory occupied by the barren soil and cities decreasing in Kolwezi, and the Basse-Kando, while it increases in Tenke-Fungurume up to almost 10% in 1990. The following decade was marked by an increase in the territory occupied by the barren soil and cities; it reached 50% in Kolwezi, Tenke-Fungurume 20%, and 10% in the Basse-Kando reserve. During the last decade, the barren soil and cities kept evolving and grow reached almost 30% in Tenke-Fungurume, and 20% in the BasseKando reserve.

The last coverage type is represented by burnt areas; the Figures 5(a) and (b), shows the evolution over the last 30 years. In 1980, the burnt areas accounted for over 10% of Kolwezi territory, 25% of Tenke-Fungurume, area: (a) Kolwezi; (b) TenkeFungurume; (c) Basse-Kando. and more than 35% of the Basse-Kando reserve. This area saw its territory decrease gradually; in 2000 it's stand for less than 1% on three study sites. The last decade was marked by the increase of the territory occupied by the burnt areas, more than 10% in Kolwezi, a little more than 5% in the Basse-Kando reserve and Tenke-Fungurume.

Table 1: Estimation of land cover type (% and ha) in the study area for 1980's, 1990's, 2000's, and 2011 years

	Landsat MSS 1980		Landsat TM 1990		Landsat ETM 2000		Spot 2010		Delta 1980-2000
Class	1980		1990		2000		2011		31
Kolwezi	%	ha	%	ha	%	ha	%	ha	
Barren soil and cities	39.3	29082.00	35.66	26388.40	48.90	36191.29	39.55	29267.00	-185.00
Burned area	12.98	9605.20	4.90	3626.36	0.84	623.22	10.80	7992.00	1613.20
Vegetation	38.64	28593.60	49.37	36537.45	40.40	29899.73	42.66	31568.40	-2974.80
Water	9.07	6711.80	10.06	7444.40	9.84	7285.74	6.99	5172.60	1539.20
Tenke-Fungurume									
Barren soil and cities	6.12	3614.54	10.77	6359.13	20.25	11951.02	27.80	16402.00	-12787.46
Burned area	26.59	15690.66	4.06	2396.54	0.16	95.13	5.33	3144.70	12545.97

Vegetation	67.27	39694.78	85.09	50208.72	78.68	46424.66	66.78	39400.20	294.59
Water	0.00	0.00	0.06	35.59	0.89	529.17	0.09	53.10	-53.10
Basse-Kando									
Barren soil and cities	6.30	2142.00	3.40	1156.00	9.20	3128.00	19.87	6755.80	-4613.80
Burned area	37.05	12597.00	10.26	3488.40	0.54	183.60	5.87	1995.80	10601.20
Vegetation	48.93	16636.20	79.87	27155.80	81.03	27550.20	68.54	23303.60	-6667.40
Water	5.71	1941.40	5.72	1944.80	5.71	194140	5.72	1944.80	-3.40

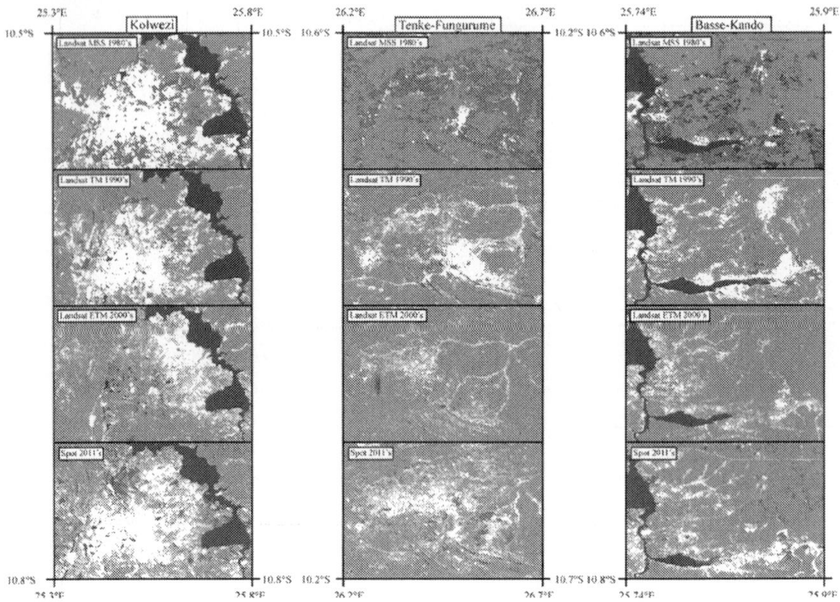

Figure 4: Temporal and spatial variation maps of the land cover types in the study area for the years: Landsat MSS (1980's), TM (1990's), ETM (2000's), and Spot (2011). Landsat classifications: the blue color stands for water, the white color for barren soil and cities, the red for burned area, and the green color for vegetation.

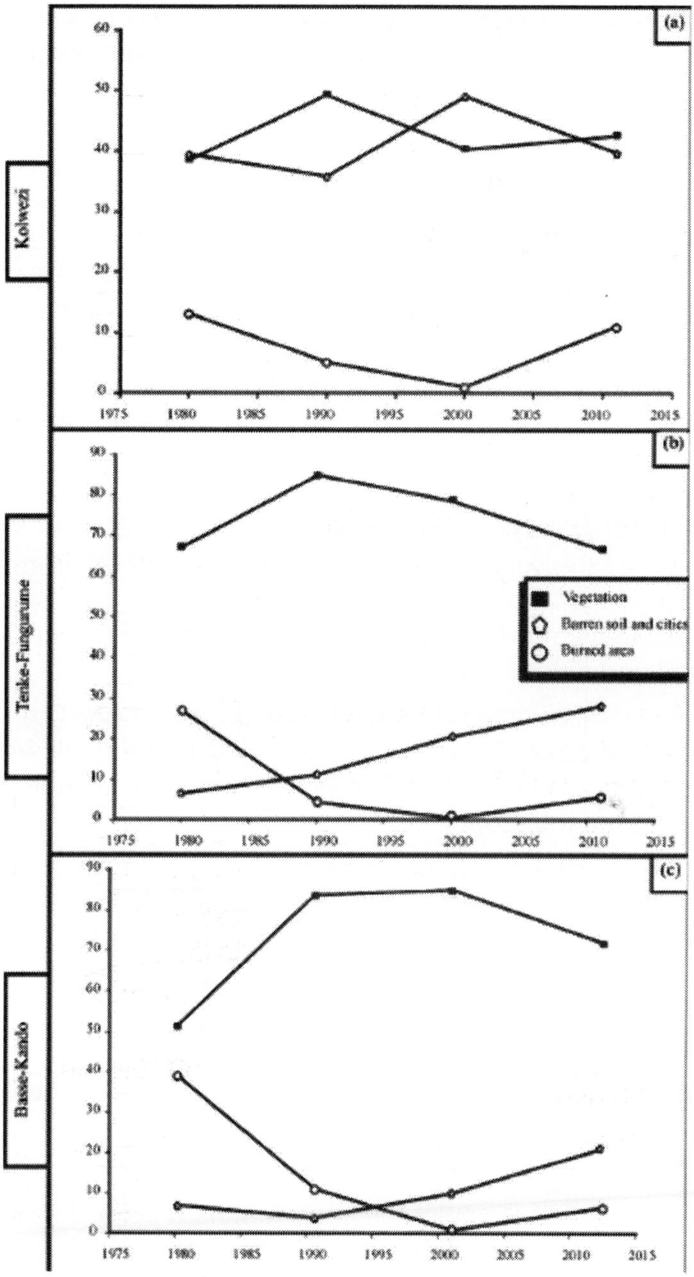

Figure 5: Temporal changes estimated in % for vegetation, barren soil and cities, and burned

The evolution dynamics of each coverage type over the last 30 years are brought in Figures 5(a)-(c). The figure suggests a different evolution for vegetation and barren soil and cities in the three study sites rather than burned areas follow a similar trend in the three study sites. These different dynamic behaviors are the focus of the following analyses.

Change in Vegetation Patch Number

The variation of patch number in the Kolwezi area is illustrated on Figure 6 and Table 2. This figure shows that the vegetation patch number increase between 1980 and 1990 is related to an increase of the vegetation area and the decrease of the barren soil and cities. This reduction is also marked by an increase of the barren soil and cities patch number. This parceling of barren soil and cities continues between 1990 and 2000, but the total area also increases leading to a reduction of the vegetation area and patch number. From 2000 to 2011, both trends are again evolving in the opposite directions with an increase in the number of vegetation patches, and a decrease in the barren soil and cities class.

The Tenke-Fugurume site shows a cyclic variation (Figure 6 and Table 2) with a peak in 1990 and a low point in 2000. However over the last 30 years the evolution curves of the patches number intersect twice in 1985 and in 2000.

Evolution curves of the patch number in the BasseKando site are drawn on Figure 6 and Table 2. These curves follow the same trend as those of Tenke-Fungurume, reaching a high level in 1990 and 2011 and lowest one in 2000. Unlike the other sites, the curves in the Basse-Kando site do not intersect.

Spatial Relation between Vegetation Cover and Barren Soil and Cities

Index of aggregation is one of the important parameters in the analysis of the land cover change. This index has been calculated on two types of land cover: vegetation, on the one hand, and barren soil and cities classes on the other hand. The change in this index in the Kolwezi area is illustrated inFigure 6; it shows the same close relationship than the previous parameters, as it involves a decrease until 1990. During the next decade, while the index of aggregation of the barren soil and cities continues to decrease, those vegetation increases, producing the first crossing in 1993 (which corresponds to the wave of migration to the east of the DRC population from neighboring Rwanda). From 2000 to 2011, the evolution of this index on the two types of land cover is evolving in the opposite direction with an increase in the index of aggregation of the barren soil and cities, and a decrease in that of vegetation, producing the second crossing in 2007 (corresponds to the resumption of fighting in eastern of DRC).

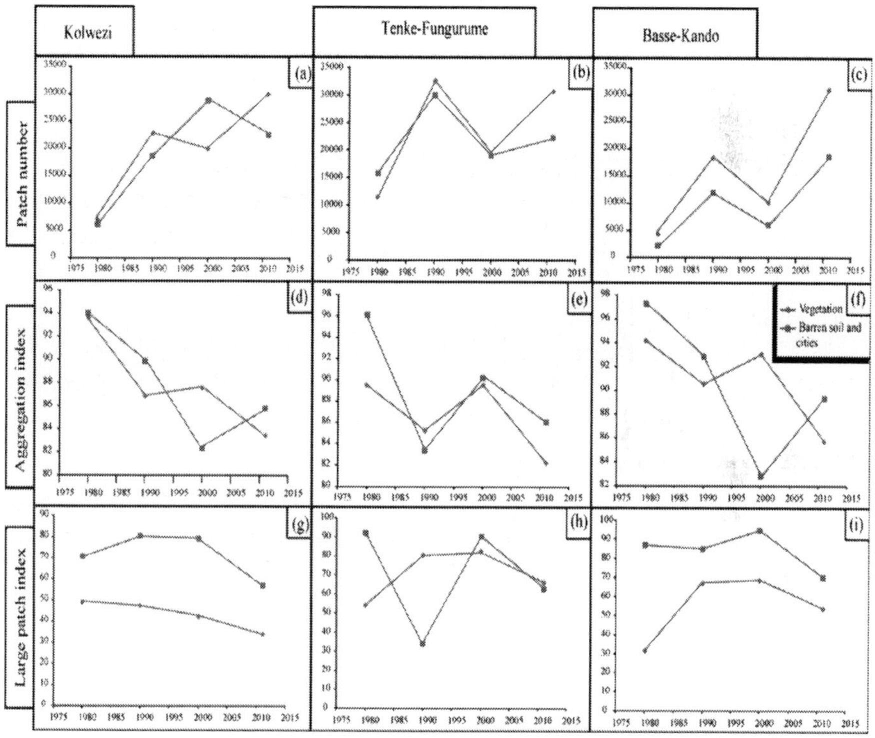

Figure 6: Temporal changes in the study area estimated in term of: %, patch number, aggregation index, and large patch index.

Table 2: Changes in landscape pattern indices for vegetation, and barren soil and cities in the study area for 1979, 1990, 2000, and 2011 years

Kolwezi		Landsat MSS 1980	Landsat TM 1990	Landsat ETM 2000	Spot 2011
Pattern indices	Vegetation	1980	1990	2000	2011
Aggregation index		93.58	86.85	87.54	83.42
Large patch index		48.87	47.33	42.18	33.96
Number of patch		6979	22850	20010	29999
Aggregation index	Barren soil	93.97	89.82	82.29	85.76
Large patch index	and cities	71.01	80.61	79.12	57.10
Number of patch		6120	18588	28894	22473
Tenke-Fungurume		Landsat MSS 1980	Landsat TM 1990	Landsat ETM 2000	Spot 2011
Pattern indices	Vegetation	1980	1990	2000	2011
Aggregation index		89.50	85.23	89.49	82.22
Large patch index		54.11	80.06	82.06	65.82
Number of patch		11438	32707	19340	30568

Aggregation index	Barren soil	96.1411	83.4281	90.2336	86.02

Large patch index	and cities	91.8266	33.9652	90.5222	62.75
Number of patch		15800	29999	19007	22159
Basse-Kando		Landsat MSS 1980	Landsat TM 1990	Landsat ETM 2000	Spot 2011
Pattern indices	Vegetation	1980	1990	2000	2011
Aggregation index		94.17	90.50	93.05	85.75
Large patch index		31.64	67.07	68.57	53.63
Number of patch		4352	18334	10120	30917
Aggregation index	Barren soil	97.23	92.84	82.80	89.29
Large patch index	and cities	87.24	84.90	94.95	69.85
Number of patch		2314	12032	6009	18623

This aggregation index (Figure 6 and Table 2) have a cyclic variation on the Tenke-Fugurume area. This variation is marked by a decrease during the first and the last decade, and an increasing curve in the second decade. The highest points correspond to the years 1980 and 2000, and the lowest points respectively to 1990 and 2011. The evolution of curves over the past 30 years shows two junctions in 1987, and in 1997.

The aggregation index curves on the site of the BasseKando are drawn on Figure 6. In opposite to the previous settings, these curves have a complementary development similar to that observed in Kolwezi. This evolution is marked by a decrease in this index until 1990. Over the next decade, while the index of aggregation of the barren soil and cities continues to decrease, those vegetation increases, producing the first crossing in 1993. During the past decade, the two curves are evolving in the opposite direction with an increase in the index of aggregation of the barren soil and cities, and a decrease in that of vegetation, producing the second crossing in 2007.

DISCUSSION

The present study constitutes an extensive analysis of land cover fragmentation conducted in Katanga. The results demonstrate how changes in spatial patterns of the land cover may be assessed using multi temporal satellite data, as has been achieved recently in other parts of the world [4,30,31].

Assessment of Methods and Results

As satellite imagery classifications into land cover types are never completely accurate, landscape pattern analyze of the present study are affected by errors in the classification which alter the accuracy. Image accuracy tends to improve as the image date becomes more recent. The supervised classification, which was strongly supported by ground-based information, provided a suitable identification of land cover types in each of the satellite scenes processed. Although the classification conducted for the oldest scene (MSS 1980) suffered from the disadvantage of limited ground validation, a cautious interpretation considering field control points that had not changed over time involving high values of accuracy to be obtained, and enabling clearly identifiable categories of land cover in the classification.

In the present study, the error was minimized by applying improved topographic correction algorithms in the image processing and by aggregating some of the subcategories of land cover types. Additionally, different ground-based surveys conducted from previous works and additional field visits were used to support the image classifications. Also, the error was minimized by filtering the classified image to remove small patches of less than four 50 m-pixels (equivalent to 0.60 ha).

Land Cover Change and Spatial Analyzis

Landscape pattern indices provide a useful tool to explore inside variability, cross-site differences and changes over time. The simultaneous use of class-level and patch-level landscape pattern indices enable assessment of the spatial configuration of vegetation cover and its relation to other land cover types.

The various indices and parameters used in the analysis of the land cover change (fragmentation) at Kolwezi site are shown on Figure 6. During the first period (1980-1990), for the vegetation, there is an increase in the patches number, but a regression of the aggregation and the large patch index. This evolution of the parameters suggests that vegetation has enlarged by the formation of small scattered patches. This period corresponds to the decrease of area occupied by barren soil and cities, despite the increase of their patch number, and the large patch index. This variation of parameters shows the regrouping of barren soil and cities, which may correspond to a concentration of the population in areas. At the same time the burned areas decreased strongly, this means a neglect of agriculture by the population that turns towards another socio-economic activity.

The next decade (1990-2000) is marked by the reduction of space occupied by vegetation, the burnt areas, and especially by a growth of barren soil and cities. This evolution is marked for the barren soil and cities by the increase in the large patch index, and the number of patch. For the vegetation, only the aggregation index increases while others parameters decrease. This variation shows that vegetation is replaced by barren soil and cities, mainly by occupation of small scattered areas. This situation would correspond to the degradation of the vegetation zones for a socio-economic activity, which in this case corresponds to the exploitation of mineral resources.

The last decade (2000-2011) is marked by the regeneration of the vegetation, and a decrease in the barren soil and cities surface. This is illustrated by the increase in the number of patches for the vegetation, and a reverse evolution for the barren soil and cities. The simultaneous decrease in the large patch index for both types of land cover illustrates the fact that vegetation has expanded by small patches, and that the barren soil and cities are small isolated areas and increasingly more consolidated.

The various indices and parameters used for the fragmentation of the land cover analysis are represented on Figure 6 for the Tenke-Fungurume site. The first decade here is marked by the regeneration of the vegetation and by the increase in the barren soil and cities. This is illustrated by an increase in the number of patches, but a regression of the index of aggregation. The large patch index brings out a different evolution; it decreased strongly for barren soil and cities and increased for vegetation. This evolution of the parameters shows that vegetation is enlarged by training large scattered patches, and so that the barren soil and cities was enlarged by the formation of small patches. During the same period, the burnt areas decrease strongly; this means a neglect of agriculture by the population that turns to another socioeconomic activity.

The second decade is marked by the reduction of the area covered by vegetation, the burnt areas, and especially by a growth of urban areas. Because of the increase in the large patch index of barren soil and cities, the aggregation index, and the decrease of other parameters, it shows a growth of barren soil and cities specifically located around existing sites. This variation suggests that vegetation and burned areas are replaced by barren soil and cities. This situation is similar to that encountered in Kolwezi where agricultural activity is depreciated in favor of a socio-economic activity, which in this case corresponds to the exploitation of mineral resources.

Over the last decade, the vegetation surface continues to decline, while the barren soil and cities, and burned areas increase. This is illustrated by a raise in the number patches, and a drop decrease the large patch index and index of aggregation for both types of ground covers. This situation brings out the constant development of the barren soil and cities involving the vegetation fragmentation in small patches. This phenomenon suggests a resumption of agricultural activity, coupled with other socio-economic activities.

The variation of the different parameters for the analysis of the fragmentation of the land cover in the BasseKando reserve is illustrated on Figure 6. The first decade was marked by an increase of the vegetation, and a decrease of barren soil and cities, and burned areas. It comes along with a raise in the patches number, and a decrease in the aggregation index for both types of land cover. The large patch index shows a different evolution with an increase for vegetation and a decrease for the barren soil and cities. This reflects an increase of vegetation by large areas close to each other and interconnected. Barren soil and cities are mainly small disconnected groups. At the same time the burned areas drop sharply meaning a neglect of agriculture by the population that turns towards another socio-economic activity.

The second decade is marked by an increase of the barren soil and cities surface, stagnation of vegetation areas, and a reduction of burnt areas occupation. This development is marked by an increase in the large patch index of barren soil and cities and vegetation, a decrease in the patch number. Only the aggregation index shows a different variation, a growth for the vegetation, and a decrease for barren soil and cities. It suggests that vegetation here is protected, and that the surface replacement is operated between the barren soil and cities and the burned areas.

During the last decade a growth of barren soil and cities, areas burned and a small decrease in vegetation is observed. It is illustrated by an increase in the patch number, for both types of land cover, a decline in the large patch index. The aggregation index shows a development in opposite directions, an increase for the barren soil and barren soil and cities and a decrease for vegetation. This situation represents a growth of barren soil and cities by small interconnected patches, and a decrease of the vegetation fragmented into small patches.

Impacts

Over 12,700 protected areas have been established around the world, accounting for 8.81% of Earth's land surface. Protected areas are a key strategy used to conserve biodiversity from human activities. Remote sensing is being used to determine how successful protected areas are at mitigating human impact [32].

The 1970's classification brings out a significant observation with the presence of very large burned patches specifically located around human activity points. The Basse-Kando hunting reserve is well-established as its creation dates back from 1957 by the Congolese government and was recognized and integrated into the Congolese mining code official journal, special number, April 2003 [33]. However, the deforestation and fragmentation rates within the reserve remained over the 30 years of the study. The vegetation regression rate remains lower in the protected (Basse-Kando reserve) area than in mine areas (Kolwezi and

Tenke-Fungurume). By adding geological data layers in the Geographic Information System (GIS) to the classification information, it allows us to distinguish the original drivers inducing vegetation regression and fragmentation in the Basse-Kando reserve. Mining companies are focusing on a highly investmentworthy economic geologic unit named the "Mines series" to exploit the copper and cobalt ores (Figure 2). The international economical context is really propitious for the exploitation of copper and cobalt ores. Indeed, the superposition of both layers indicates that the barren land patches within the mine series unit are generated by mining activities and the patches outside are human settlements. The encouraging economic copper and cobalt consumption trends for the forthcoming years imply that the pressure on the Basse-Kando reserve is going to increase and the fragmentation rate will raise in the "mines series" specific location zones. The same trend is observed at the international scale in [34].

The mining areas of Tenke-Fungurume and Kolwezi in the buffer conservation zones as well as the Basse-Kando reserve in Katanga, South-eastern (DRC), are being studied through analysis of remotely sensed data for evaluation of impact due to mining and related activities on the environment. The comparison between different industrialization periods covering a time span of about 30 years depicted through remotely sensed reveals that the face of the assessed areas changed progressively with the development of mining activities. A detailed assessment of impact of mining changes in LULC pattern and fragmentation on time and space has been undertaken to illustrate and quantify the changes. By means of multispectral and multi-date remote sensing data processing and classification, the results show 1) an evolution from extensive large to small scale denudation patches can be observed from 1970's to 2000's; 2) 3 observation types corresponding to vegetation cover regression in the 1970's, vegetation reclamation in the 1990's, then again a vegetation regression period in the 2000's and going on more intensively until 2011. Moreover, the comparison of land cover and geological maps and protected areas clearly shows that mining sites are increasingly overlapping vegetation. It brings out a similar trend of vegetation pattern evolution as the buffer areas but not at the same intensity.

The classification of the Basse-Kando reserve, the Kolwezi and Tenke-Fungurume mining centres from the 1979 until 2011 shows different trends, with vegetation regeneration between 1979's and 1990's followed by regression phase since the 1990's.

The observed vegetation changes in the vicinity of the mining sites are the results of (1) direct factors such as the evolution in the type of mining extraction and the variation of the ore production volumes and also (2) to factors indirectly derived from the mining activities such as demographic influx due to the attraction of job offers, economic growth, agriculture development (Figure 2). Deforestation became the dominant process, involving a decline in both mean patch number. The process of fragmentation comes along with the loss of the largest vegetation patch (Table 2). The gradual increase to mining and non-vegetation areas revealed the pressure of this activity on the surrounded landscape. However, most of the mining concessions and operational mining sites are located in governmental protected areas and surrounded protected buffer zones (Figure 2).

Potentials Drivers

The presence of large burnt and barren soil areas is largely specific to the 1970's as no such event was observed in the other classification maps. The field campaign interviews gave the opportunity to get some information about this time and to identify the drivers, which triggered the large fire episodes in the area. Indeed this period corresponds to governmental investments to relaunch and stabilizes the DRC economy after the independence of the country which led to some economic instability.

At the beginning of the 1980's, Kolwezi and TenkeFungurume area some of the spots which benefited from investments in new ore deposit exploitation. Intensive exploration campaigns were carried out in those areas with drilling and trenching programmes as well as copper clearings on surrounding hills, feasibility studies and infrastructure building (airport, buildings for administration and workers, road access, concentrator, etc). These exploration activity appears later in the Basse-Kando area. The burning signature is then considered to be from an anthropogenic origin. Thus, in this study, this signature is associated to the barren soil class in the assessment of mining operation impact. The 1980's classification brings out a significant observation with the presence of very large burned patches specifically located around human activity points.

CONCLUSIONS

In this paper, we demonstrate capacity to monitor the vegetation status versus the mining activities of a natural reserve and two important mining areas using a combination of Geographic Information System (GIS), remote sensing techniques, field-based observations and textual data resources. The multi-temporal remote sensing approach is used in the assessment of the spatial changes, their organisation and the determination of the agents involved in the land cover fragmentation. This information coupled with geological data, parks and reserves boundaries suggests the trend of the future pressure on land cover in the study area. The investigation goes further with the identification of drivers which are essential to understand the processes of changes.

The structure of the study areas catchment has changed significantly during the study year period. This study quantifies the degenerating condition of the land cover due to anthropogenic activities. The land cover areas have become more fragmented and are characterized by the proliferation of much smaller, less connected patches. The management of those areas, particularly the identification and prioritization, rehabilitation, preservation, and the development of supporting policies, could be expected to benefit from the use of landscape structure information. Moreover, no environmental assessment is available in the Katanga province, which reveals a real lack in environmental management [15]. Scientific mining advices should provide alternative operational methods for a better use of natural mineral resources by minimizing the damages. Proactive management with a focus on conservation and sustainable management are imperative.

ACKNOWLEDGEMENTS

The authors wish to thank the GECO program (Geology for an ECOnomic sustainable development) financed by the Foreign Affairs Ministry.

REFERENCES

1. J. A. Foley, R. DeFries, G. P. Asner, C. Barford, G. Bonan, R. S. Carpenter, F. S. Chapin, M. T. Coe, G. C. Daily, H. K. Gibbs, J. H. Helkowski, T. Holloway, E. A. Howard, C. J. Kucharik, C. Monfreda, J. A. Patz, I. C. Prentice, N. Ramankutty and P. K. Snyder, "Global Consequences of Land Use," Science, Vol. 309, No. 5734, 2005, pp. 570-574.doi:10.1126/science.1111772
2. A. Di Gregorio and L. J. M. Jansen, "New Concepts for a Land Cover Classification System," The Land, Vol. 1, 1998, pp. 55-65.
3. A. B. Brink and H. D. Eva, "Monotiring 25 Years of Land Cover Change Dynamics in Africa: A Sample Based Remote Sensing Approach," Applied Geography, Vol. 29, No. 4, 2009, pp. 501-512. doi:10.1016/j.apgeog.2008.10.004
4. J. E. Wasige, T. A. Groen, E. Smaling and V. Jetten, "Monotoring Basin-Scale Land Cover Changes in Kagera Basin of Lake Victoria Using Ancillary Data and Remote Sensing," International Journal of Applied Earth Observation and Geoinformation, Vol. 21, 2013, pp. 32-42. doi:10.1016/j.jag.2012.08.005
5. B. L. Turner II, D. Skole, S. Sanderson, G. Fischer, L. Fresco and R. Leemans, "Land-Use and Land-Cover Change; Science/Research Plan," IGBP Report No. 35, HDP Report No. 7, IGBP and HDP, Stockholm and Geneva, 1995.
6. B. L. Turner II and B. L. Meyer, "Global Land Use and Land Cover Change: An Overview," In: W. B. Meyer, B. L. Turner II, Eds., Changes in Land Use and Land Cover: A Global Perspective, Cambridge University Press, Cambridge, 1994, pp. 3-10.
7. L. J. M. Jansen and A. Di Gregorio, "Parametric Land Cover and Land-Use Classifications as Tools for Environmental Change Detection," Agriculture, Ecosystems and Environment, Vol. 91, No. 1-3, 2002, pp. 89-100. doi:10.1016/S0167-8809(01)00243-2
8. A. François, "Stratigraphy, Tectonics and Mineralization in the Copper arc of Shaba (Rep. of Zaire)," In: P. Bartholomé, Ed., Stratiform Deposits and Copper-Bearing Provinces, Centennial Geological Society of Belgium, Liege, 1974, pp. 79-101.

9. A. François, "Geological Synthesis on the Shaba Copper Arc (Rep. Zaire)," Centennial Belgian Society of Geology, 1987, pp. 15-65.
10. A. François, "Problems Relating to the Katangian of Shaba," In: M. Wendorff and L. Tack, Eds., Late Proterozoic Belts in Central Africa, Musée Royal de l'Afrique Centrale, Tervuren, 1995, pp. 1-20.
11. J. L. H. Cailteux, "Proterozoic Sediment-Hosted Base Metal Deposits of Western Gondwana," 3rd IGCP-450 Meeting and Field Workshop, Lubumbashi, 14 to 24 July 2003, p. 223.
12. M. J. Batumike, A. B. Kampunzu and J. H. Cailteux, "Petrology and Geochemistry of the Neoproterozoic Nguba and Kundelungu Groups, Katangan Supergroup, Southeast Congo: Implications for Provenance, Paleoweathering and Geotectonic Setting," Journal of African Earth Sciences, Vol. 44, No. 1, 2006, pp. 97-115. doi:10.1016/j.jafrearsci.2005.11.007
13. S. Dewaele, P. Muchez, H. El Desouky, M. Haest and W. Heijlen, "Multiphase Origin of the Base Metal Deposits in the Lufilian Fold-and-Thrust Belt, Katanga (Democratic Republic of Congo)," 2008. http://www.gecoproject.org/?
14. A. B. Kampunzu, J. L. H. Cailteux, A. F. Kamona, M. M. Intiomale and F. Melcher, "Sediment-hosted Zn-Pb-Cu Deposits in the Central African Copperbelt," Ore Geology Reviews, Vol. 35, No. 3-4, 2009, pp. 263-297. doi:10.1016/j.oregeorev.2009.02.003
15. P. Feeney, "Extractive Industries—Their Contribution to the Congos Development," Natural Resources Management Issue Paper No. 2, Center on International Cooperation, New York, 2009.
16. http://glcf.umiacs.umd.edu/index.shtml
17. P. S. Chavez, "Image-Based Atmospheric CorrectionsRevisited and Improved," Photogrammetric Engineering Remote Sensing, Vol. 62, No. 9, 1996, pp. 1025-1036.
18. T. M. Lillesand and R. W. Kiefer, "Remote Sensing and Image Interpretation," 4th Edition, John Wiley & Sons, New York, 1994, pp. 477-482.
19. M. Zhang, K. L. Carder, F. E. Muller-Karger, Z. Lee and D. B. Goldgog, "Noise Reduction and Atmospheric Correction for Coastal Applications of Landsat Thematic Mapper Imagery,"

Remote Sensing of Environment, Vol. 70, No. 2, 1999, pp. 167-180.doi:10.1016/S0034-4257(99)00031-0

20. J. J. Kashaigili, B. P. Mbilinyi, M. Mccartney and F. L. Mwanuzi, "Dyanamics of Usangu Plains Wetlands: Use of Remote Sensing and GIS as Management Decision Tools," Physics and Chemistry of the Earth, Vol. 31, No. 15-16, 2006, pp. 967-975.
21. K. Mcgarigal, S. A. Cushman, M. C. Neel and E. Ene, "Fragstats: Spatial Pattern Analysis Program for Categorical Maps," Landscape Ecology Program, University of Massachusetts, Boston, 2002.
22. D. Armenteras, F. Gast and H. Villareal, "Andean Forest Fragmentation and the Representativeness of Protected Natural Areas in the Eastern Andes, Colombia," Biological Conservation, Vol. 113, No. 2, 2003, pp. 245-256. doi:10.1016/S0006-3207(02)00359-2
23. S. Franklin, "Remote Sensing for Sustainable Forest Management," Lewis Publishers, Boca Raton, 2001. doi:10.1201/9781420032857
24. M. Fitzsimmons, "Effects of Deforestation and Reforestation on Landscape Spatial Structure in Boreal Saskatchewan, Canada," Forest Ecology and Management, Vol. 174, No. 1-3, 2003, pp. 577-592. doi:10.1016/S0378-1127(02)00067-1
25. J. Imbernon and A. Branthomme, "Characterization of Landscape Patterns of Deforestation in Tropical Rain Forests," International Journal of Remote Sensing, Vol. 22, No. 9, 2001, pp. 1753-1765.
26. A. C. Millington, X. M. Velez-Liendo and A. V. Bradley, "Scale Dependence in Multitemporal Mapping of Forest Fragmentation in Bolivia: Implications for Explaining Temporal Trends in Landscape Ecology and Applications to Biodiversity Conservation," ISPRS Journal of Photogrammetry and Remote Sensing, Vol. 57, No. 4, 2003, pp. 289-299. doi:10.1016/S0924-2716(02)00154-5
27. N. Staus, J. Strittholt, D. Dellasala and R. Robinson, "Rate and Patterns of Forest Disturbance in the KlamathSiskiyou Ecoregion, USA between 1972 and 1992," Landscape Ecology, Vol. 17, No. 5, 2002, pp. 455-470. doi:10.1023/A:1021274701133
28. M. Steininger, C. Tucker, P. Ersts, T. Killeen, Z. Villegas and S. Hecht, "Clearance and Fragmentation of Tropical Deciduous Forest in the Tierras Bajas, Santa Cruz, Bolivia," Conservation

Biology, Vol. 15, No. 4, 2001, pp. 856-866. doi:10.1046/j.1523-1739.2001.015004856.x

29. C. Echeverria, D. Coomes, J. Sals, J. M. Rey-Benayas, A. Lara and A. Newton, "Rapid Deforestation and Fragmentation of Chilean Temperate Forests," Biological Conservation, Vol. 130, No. 4, 2006, pp. 481-494. doi:10.1016/j.biocon.2006.01.017
30. H. Tommervik, K. Hogda and I. Solheim, "Monitoring Vegetation Changes in Pasvik (Norway) and Pechenga in Kola Peninsula (Russia) Using Multitemporal Landsat MSS/TM Data," Remote Sensing of Environment, Vol. 85, No. 3, 2003, pp. 370-388. doi:10.1016/S0034-4257(03)00014-2
31. W. Rees, M. Williams and P. Vitebsky, "Mapping Land Cover Change in a Reindeer Herding Area of the Russian Artic Using Landsat TM and ETM+ Imagery and Indigenous Knowledge," Remote Sensing of Environment, Vol. 85, No. 4, 2003, pp. 441-452.doi:10.1016/S0034-4257(03)00037-3
32. A. A. Apan, S. R. Raine and M. R. Paterson, "Mapping and Analysis of Changes in the Riparian Landscape Structure of the Lockyer Valley Catchment, Queensland, Australia," Landscape and Urban Planning, Vol. 59, No. 1, 2002, pp. 43-57. doi:10.1016/S0169-2046(01)00246-8
33. Mining code of the Democratic Republic of the Congo. "Law No. 007/2002 of the 11th July, 2002 Relating to the Mining Code," Official Journal Special No. of the 15th July 2002.
34. G. O. Obati and G. Owuor, "Extractive Industries, Natural Resources Management and Sustainable Development: A Review," Affiliated Network for Social Accountability, ANSA Africa, Pretoria, 2010.

Chapter 2

Land Cover Transformation in Two Post-Mining Landscapes Subjected to Different Ages of Reclamation since Dumping of Spoils

Effah K Antwi[1, 2], John Boakye-Danquah[3], Stephen B Asabere[4], Kazuhiko Takeuchi[1, 2], and Gerhard Wiegleb[5]

[1]United Nations University- Institute for the Advanced Study of Sustainability (UNU-IAS), Tokyo, Japan

[2]The University of Tokyo, Todai Institutes for Advanced Study, Integrated Research System for Sustainability Science (IR3S), Tokyo, Japan

[3]School of Environment and Sustainability, University of Saskatchewan, Saskatoon SK S7N 5C8, Canada

[4]Faculty of Environmental Sciences, Department of Geosceinces, Technische Universitaet Dresden, Dresden, Germany

[5]Environmental Science and Process Engineering, Department of General Ecology, Technical University of Brandenburg (BTU), Cottbus, Germany

ABSTRACT

Transformation of natural land cover (LC) into modified LC has become inevitable due to growing human needs. Nevertheless, landscape transformational patterns during reclamation of mine damaged lands remain vague. Our hypothesis was that post-mining landscapes with different ages since dumping become more diverse in LC transformation over time. The aim was to study the impact of landscape reclamation on land cover changes (LCC) in two post-mining landscapes. Land cover maps of 1988, 1991, 1995, 1998, 2000 and 2003 were produced from LANDSAT TM images of Schlabendorf Nord and Schlabendorf Süd and used to survey the changing landscape. Change detection extension was used to identify changes among land cover types (LCTs). Detrended correspondence analyses (DCA) ordination technique (CANOCO) aided study of similarity among LC distribution. Soil pH analysis was carried out to study effect of soil and climate conditions on LCC. The results show that visible patterns of increase and decrease in the LCTs occurred in both landscapes. Given two post-mining landscapes subjected to different ages of reclamation, clear differences in vegetation growth and LCC pattern would occur. At early stages of restoration, LCTs often have unstable conditions and experience more acute transformation depending on the level of land use intensity in space and time. LCCs were mostly due to progressive and reversed succession. Due to variation in post-mining landscape soil conditions, soil treatment during reclamation should be site specific. The comparative analysis of LCCs in Schlabendorf provides a framework for prioritizing land use planning options for sustainable management of post-mining landscapes in temperate ecosystems.

INTRODUCTION

Anthropogenic disturbance such as surface coal mining is particularly rigorous in nature. Regardless of the method used and the resources available, it affects land cover (LC) at the site of extraction (Haigh 2000); leaving behind multiple damages that stretch over a wide range of land. Continuous monitoring of mining induced land cover change (LCC) is essential for sustainable mine management. Particularly, it helps to identify the impact of mining on land cover types (LCTs) and to provide the needed mine closure and reclamation measures (Demirel et al. 2011).

Ecosystems undergo long periods of natural succession and it could take hundreds of years for full recovery to occur, if it does occur at all. Human mediated restoration measures are often needed to accelerate succession processes (Groninger et al. 2007; Hendrychová 2008). Reclamation of mine-damaged lands is feasible but it may be a slow or comparatively fast process depending on the extent of damage. Therefore reclamation schemes that consider some kind of human intervention could be adapted to accelerate restoration processes and reinstate the ecological integrity of damaged landscape, giving it a defined after use. Bradshaw (1999) reported a successful study by Western Australia mining company, Alcoa, who stockpiled and reused the topsoil to restore an area after mining. Knoche et al. (2000) ascertained that ecosystems in Lower Lusatia gradually shift from a state of geochemical-dominated process to a mode of increasing biological control.

Haigh (2000) ascertained that high acidity and high concentration of aluminums in coal-mining areas are often responsible for land degradation (loss of vegetation cover). Post-mining reclamation schemes are ultimately done to help degraded lands to be productive again; and this mainly involves re-establishment of vegetation cover. For successful post-mining landscape reclamation, sustainable practices such as management and replacement of top soil should be vital during moving and storage phases of excavation (Nova Scotia Environment 2009). This eventually provides the condition favorable for the establishment of a self-sustaining, succession-based vegetation cover through the application of native seed mixtures and plant materials (Nova Scotia Environment 2009).

One of the main goals of the state of Brandenburg government in the Lusatia post-mining landscape reclamation was to enhance the production of biomass for energy use (obtain a share of 3% of the primary energy consumption from biomass) on arable set-aside and post-mining land (Bungart and Hüttl 2001). Short-rotation plantations were planted as energy forests that served the purpose of bioenergy or wood fuel production on recultivated land.

In landscape ecology, vegetation is considered as mosaic of patches with unique landforms, species composition and disturbance gradient

(Ravan et al. 1998; Antwi et al. 2008). Visualizing and analyzing the landscape mosaic requires technological innovations which are mainly found in geospatial technologies including multispectral Remote Sensing (RS) and Geographic Information Systems (GIS). Launching of the LANDSAT sensor in 1972 aided the application of remote sensing for estimating terrestrial plant primary production based on Normalized Differential Vegetation Index (Young and Harris 2005). It is now possible to depict the assemblage of dominant growth forms of plant species with common habitat to identify vegetation and LC units which are essentially missing during field survey due to limitations in sampling technique (Latifovica et al.2005).

Ravan et al. (1998) used GIS to map out disturbance zones in natural ecosystems to investigate its impact on primary production and biodiversity. Gould (2000) applied remote sensing to detect patterns of vegetation and variations in biodiversity at the mesoscale. At different spatial scales, it has become possible to map the discontinuous distribution LCTs by using LANDSAT in GIS environment (Edenius et al. 2003; Cohen and Goward 2004).

LC distribution may differ or show similarity due to the land use (LU) intensity, the age of LCT and the type of LC. The level of similarity can be monitored using detrended correspondence analysis (DCA). DCA gives the best performance among all tested ordination techniques and provides effective approximation solutions to ordination problems (Hill and Gauch 1980; Palmer 2010). Ter Braak (1996) found that in DCA, sites are represented by points and each site is located at the center of gravity of the species that occur there.

The importance of temporal and spatial LU transformations has been addressed by many LU and LCC studies (Swangjang and Iamaram 2011). LCC detection has become a central component in current strategies for monitoring LU and environmental changes (Latifovica et al. 2005). It provides better understanding of landscape dynamics during a known period of time. Through the use of spatial and temporal analyses techniques from GIS applications, change detection can provide useful applications in a wide variety of areas such as LUCs, habitat structure etc. (Latifovica et al. 2005; Rahdary et al. 2008; Antwi et al. 2014). Yang (2008) applied a satellite and airborne remote sensing technique to extract LC information that depicts present and potential landscape dynamic and structure analysis. In order to quantify the anthropogenic causes of LCC in the Lake Tahoe basin resulting from varying LU intensity on forest landscapes, Raumann and Cablk (2008), with the aid of post-classification change analysis, studied changes in four broad historical periods. They ascertained that decreased area of forest, wetlands, and shrub' degradation corresponded with increase in built-up areas.

The research approach adopted in this study offers a trade-off between expensive ground-based LCC survey and cost effective satellite-based data analysis. The general aim of the research is to compare timely LC transformation in two post-mining landscapes which have been subjected to different ages of reclamation activities using GIS and remote sensing. The specific objectives of the research are to:

- Identify and quantify the area of all LCTs in the two post-mining landscapes to determine the impact of surface coal mining on LC transformations and vegetation growth in the study area.
- Study changes and similarities in LC transformation trends in two post-mining landscapes which have been subjected to different ages of land reclamation since dumping of spoils.
- Study the interactive distribution of different habitat types and determine their relationship with environmental conditions.

The following questions were answered:

- How does spatial and temporal LCC relate to changing environmental conditions?
- Are the post-mining landscapes getting more diverse over time?

MATERIALS AND METHODS

Study Area

The study area, Schlabendorf, is located in the Lower Lusatia region of Germany. It is regarded as an extremely disturbed post-mining landscape. Coal mining in Lusatia devastated more than 80,000 ha of land surface. However, more than 250,000 ha of the region was affected by the lowering of ground water due to extensive pumping measures prior to the opening of the mine, resulting in a groundwater deficit (7×109 m^3) in the region (Hüttl 1998).

The two post-mining landscapes are named after the village Schlabendorf (Figure 1); located between both post-mining landscapes at latitude 51.817 and longitude 13.817; dividing it into North (Nord) and South (Süd). Mining activity in Schlabendorf started in 1975 and ended 1991. Schlabendorf Süd has undergone 19 years of reclamation since dumping, and covers 3,035 ha of land surface. Schlabendorf Nord is smaller (2,585 ha), with 35 years of reclamation. Spoils in the post-mining landscape have dry habitat condition (Schultz and Wiegleb 2000) characterized by acidic sandy spoils on lignite rich tertiary sand. They have low water retention capacity and high evaporation rate high carbon content, low water capacity, low nutrient content and lack of biological activity (Grünewald 2001).

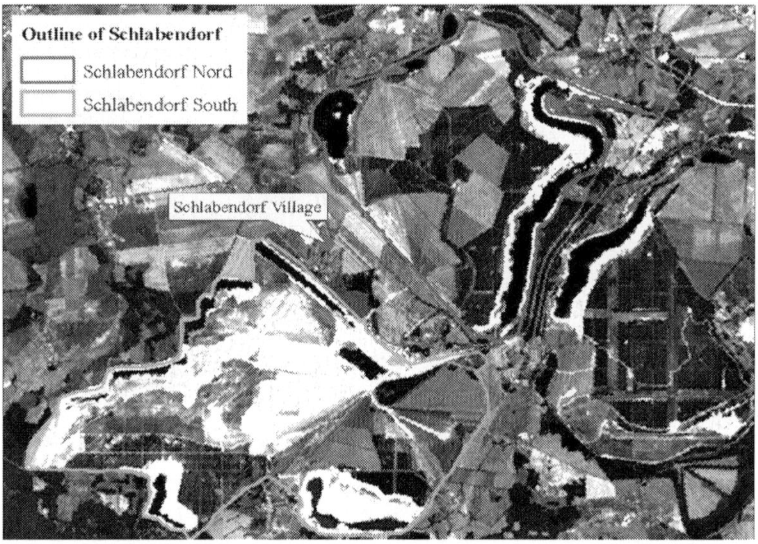

Figure 1: Outlines of Schlabendorf Nord and Schlabendorf Süd on a LANDSAT TM image of 2000.

Extensive studies on the ecological and socioeconomic problems, dominant vegetation types and ages are given by Blumrich et al. (1998), Wiegleb and Felinks (2001a, [b]), and Antwi et al. (2008). Common herbivores in the area include roe deer (*Capreolus capreolus*), wild boar (*Sus scrofa*), hare (*Lepus europaeus*), rabbit (*Lepus curpaeums*) and red deer (*Cervus elaphus*). In Figures 2 and 3 we classified the LCTs and dominant vegetation cover in both landscapes.

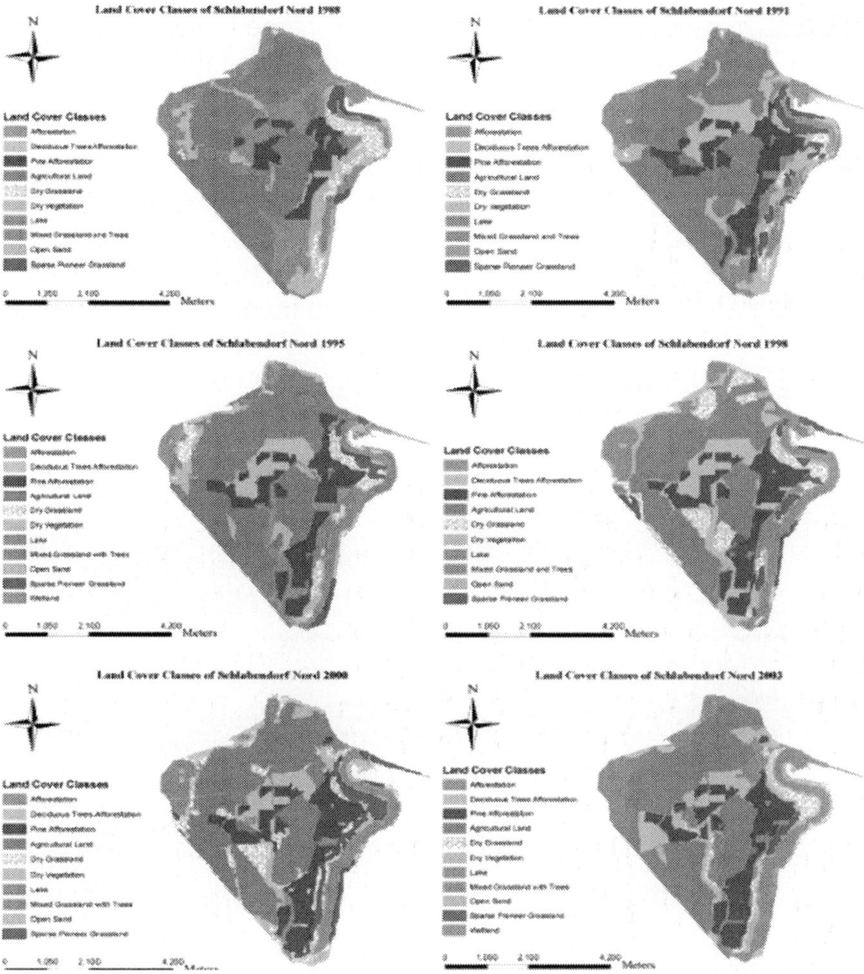

Figure 2: Final classification of land cover types in Schlabendorf Nord from 1988 to 2003 showing land cover classes with and without Wetland.

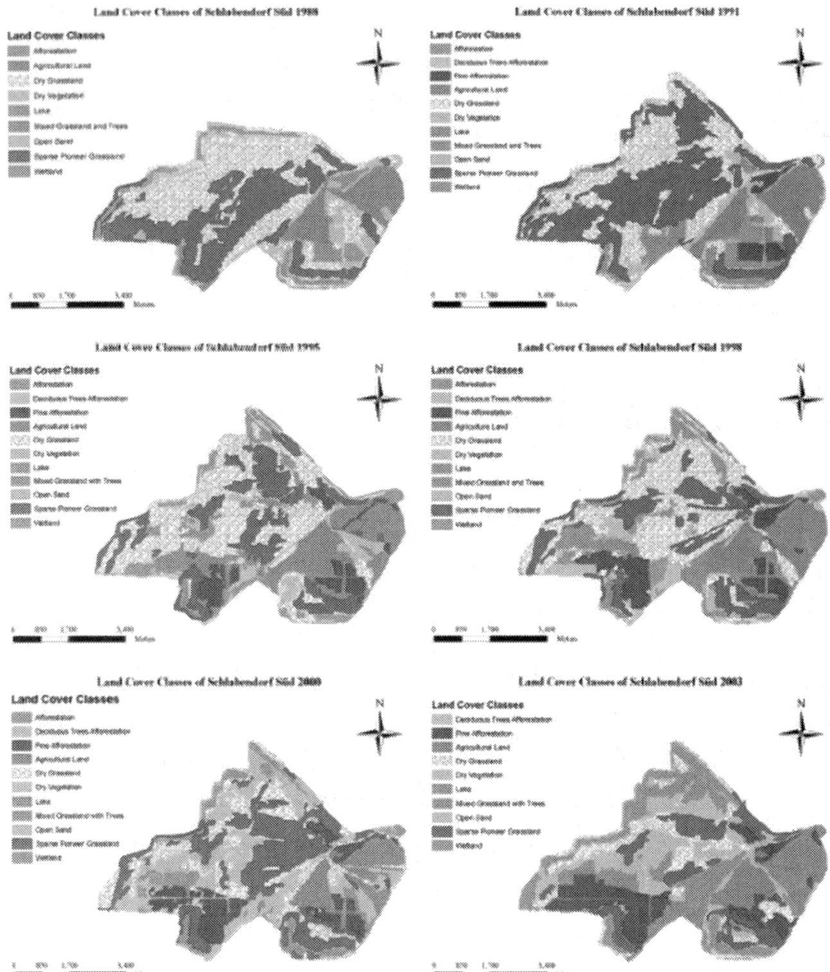

Figure 3: Final classification of land cover types in Schlabendorf Süd from 1988 to 2003 showing land cover classes with Wetland present in all periods of study.

Data Acquisition and Spatial Database

LANDSAT TM 4 and 5 and SPOT imagery of the two landscapes in the years 1988, 1991, 1995, 1998, 2000 and 2003 were used to survey the changing landscape. SPOT image was only used in 2003. Classification of LCT was based on supervised classification method using dominant vegetation type found during the years of study (Schultz and Wiegleb 2000; Antwi et al. 2008). Overall classification accuracy>95% was attained in all map years to ensure high accuracy in the final classification output (Ravan et al. 1998). For each map ear, both landscapes and LC statistics were calculated. Also, landscape metrics were computed using Patch Analyst V-3.1 (Antwi et al.2008). The scale of assessment was based on changes within habitat types in the landscape, i.e., habitat-mosaic (Weber et al. 2004; Young and Harris 2005).

Land Cover Change Detection with the Extension "Veränderung" (V3)

After the LC classification, change detection extension "Veränderung"(v3) was used to quantify changes from one LCT to another in both landscapes (Antwi et al. 2014). LCC maps were produced showing "negative change", "no change" and "positive change". All calculated changes were done using vector data model in an ArcView GIS software environment since the "Veränderung" works only on polygon files. "Veränderung" determines changes between two-polygon themes representing features of the same area. The two polygon themes intersect and form a union. The union polygons eventually become the output polygon theme. Non–intersecting areas are selected together with remaining records within polygons that are intersected by union (Antwi et al. 2008). Based on the selected intersecting and non-intersecting polygons, positive, negative and maintained areas are calculated from the attribute tables. Further analyses were made from the attribute table of the polygon containing the detected changes.

Estimation of Vegetation Growth (Primary Production) in Schlabendorf

Vegetation growth which is described in this study as primary production was calculated by greenness index using Normalized Differential Vegetation Index (NDVI). NDVI is the ratio of amount of energy reflectance in the near-infrared (NIR) and visible red portions of the electromagnetic spectrum, i.e., 0.72–1.10 and 0.58–0.68 mm respectively. The NDVI was calculated from LANDSAT TM 4 and 5 imagery (in 1988, 1991, 1995, 1998 and 2000) and SPOT imagery (in 2003) during vegetation period of both landscapes (formula 1) to investigate extent of vegetation growth during the reclamation. After masking the study area using ERDAS (ERDAS, 2002), near infrared (TM band 4 or SPOT band 3) and red (TM band 3 or SPOT band 2) were extracted. In the ArcGIS environment, spatial analyst extension was used for computation of biomass accumulation (NDVI) values as follows:

For LANDSAT TM images:

$$NDVI = (\frac{Band4 - Band31}{Band4 + Band3})100 \quad (1)$$

The resulting layer was made permanent. In instances where SPOT images were used, TM band 4 was replaced by SPOT band 3 and TM band 3 was replaced by SPOT band 2. NDVI values were then estimated from the attribute tables as shown in formula (2).

$$Biomass\ Accumulation(NDVI) = \sum(\frac{Count \times Value}{Count}) \quad (2)$$

The output NDVI layer has values ranging from -1 to +1. Higher NDVI values represent a more active growth or primary production. NDVI values below 0 represent vegetation free surface to less active vegetation growth.

The Relationship among Land Cover Types (LCTs) From Detrended Correspondence Analyses

Detrended correspondence analyses (DCA) were performed on LCT data in both landscapes for LC distribution analyses. DCA was used because it provides an effective approximation solution to the ordination problem (Hill and Gauch 1980). The CanoDraw CANOCO program (v-4.5) was used to plot the ordination diagram (ter Braak and Smilauer 2002) for Schlabendorf Süd and Schlabendorf Nord to display similarities among LC distributions in the ordination plane during the study period. Based on the position of the six study years on axis 1 of the ordination plane, the first DCA ordination axis correlates with gradient of increasing time of LCT from 1988 to 2003. Ordination axes two (2) explain 60.5% and 67.6% variance of the plot in Schlabendorf Süd and Schlabendorf Nord respectively, and they are both significant (Table 1). The second axis increases with increasing LC area. The further apart LCTs are from both ordination axes, the less the similarity in their spatial distribution in the landscape. LCTs close to the center of both ordination axes are those that did not undergo much transformation in area over time. Locations of LCTs at the edge are often unusual, but extreme pattern could lead to this occurrence, in which case additional data might be helpful for further clarification (Ter Braak 1996).

Table 1: Ordination statistics showing eigenvalues and how the variance explain the distribution of LCTs along the primary axes of DCA at Schlabendorf Süd and Schlabendorf Nord

	Schlabendorf Süd	Schlabendorf Nord		
	Axis 1	Axis 2	Axis 1	Axis 2
Eigenvalues	0.121	0.008	0.097	0.009
Variance explained (%)	56.9	60.5	61.6	67.6
Lengths of gradient	0.981	0.394	0.793	0.513

Antwi *et al.*

Antwi *et al. SpringerPlus* 2014 3:702, doi:10.1186/2193-1801-3-702

Investigating the Effect of Soil Condition on Vegetation Growth

In order to investigate reasons behind failed vegetation growth along the path of mine strip in previously afforested areas, a pseudo slash shape ( ) line transect was placed through areas with successful pine growth and failed pine growth. Thousand grams (g) of soil samples were taken from the topsoil layer at a depth of 0-10 cm (AG soil 1994). The samples were collected during the vegetation period at 0.50 m interval along the two horizontal and one vertical arms of the transect for soil acidity analyses (pH). This helped to study the influence of soil pH on vegetation growth and nutrient availability.

RESULTS

Land Cover Distribution in Schlabendorf

The LC of the study areas as observed from 1988 to 2003 was classified into eleven LC types, namely afforestation, deciduous tree afforestation, afforestation of pine trees, agricultural land, dry grassland, dry vegetation, lake, mixed grassland with trees, open sand, sparse pioneer grassland and wetland. Dominant LCTs found in Schlabendorf Nord and Schlabendorf Süd are given in Figures 2 and 3. Results of LC distribution in the two post-mining landscapes during 1988, 1991, 1995, 1998, 2000 and 2003 are grouped under five major LC categories. These are: agriculture which is made of forest lands (including afforestation, deciduous trees afforestation and afforestation of pine trees), grasslands, including dry grassland, mixed grassland and trees and sparse pioneer grassland; and water bodies made up of lakes and wetlands. The last two are minor LCTs: dry vegetation and open sand. Figure 4 gives a description of LC Distribution (in ha) in Schlabendorf Nord and Schlabendorf Süd during 1988, 1991, 1995, 1998, 2001 and 2003.

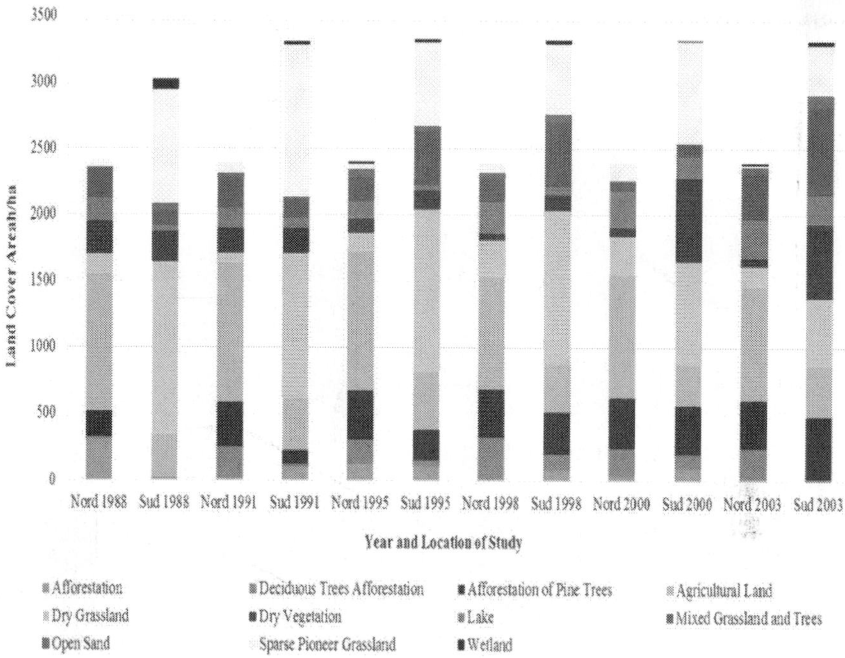

Figure 4: Description of land cover distribution (in ha) in Schlabendorf Nord and Schlabendorf Süd during 1988, 1991, 1995, 1998, 2001 and 2003.

Land Cover Distribution in Schlabendorf Süd from 1988 to 2003

Agricultural land generally increased between the study periods, though there were fluctuations in different years. In 1988, agricultural land occupied 321.97 ha and this increased to 399.18 ha and 442.18 ha in 1991 and 1995 respectively. However, land allocated for farming in 1998 decreased to 370.60 ha and again to 316.80 ha in 2000, though there was a marginal increase to 396.79 ha in 2003 (Figures 4, 5 and Table 2).

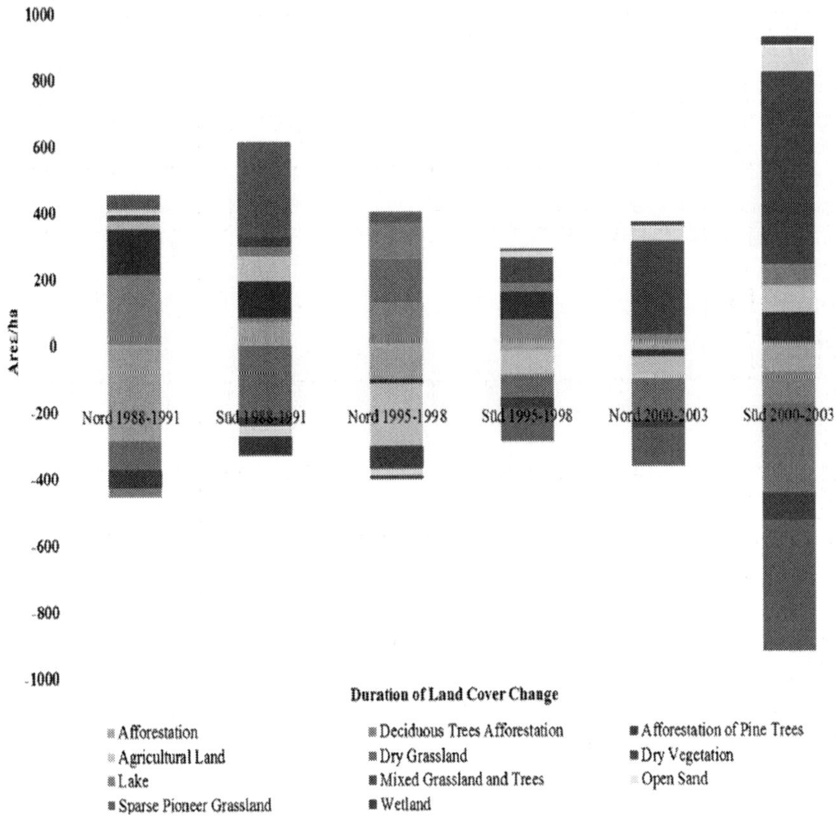

Figure 5: Increase and decrease in land cover area during 1988–1991, 1995–1998 and 2000–2003 in Schlabendorf Nord and Schlabendorf Süd.

Table 2: Land covers distribution in Schlabendorf Süd from 1988 to 2003

Land use division (Süd)	1988/ha	1991/ha	1995/ha	1998/ha	2000/ha	2003/ha
1. Agriculture						
Agricultural Land	321.97	399.18	442.18	370.6	316.8	396.79
2. Forest land						
Afforestation	25.47	96.89	105.02	84.79	87.64	0

Deciduous Trees Afforestation	0	12.67	34.47	107.89	97.77	6.27
Afforestation of Pine Trees	0	108.14	232.5	315.03	373.32	467.54
3. Grasslands						
Dry Grassland	1299.55	1084.71	1222.43	1154.39	776.28	505.81
Mixed Grassland and Trees	114.63	143.26	411.91	488.22	89.89	671.3
Sparse Pioneer Grassland	868.65	1155.21	634.37	532.73	775.1	381.13
4. Water bodies						
Lake	53.6	80.15	47.61	76.37	169.02	232.18
Wetland	72.99	19.71	17.54	24.71	0.95	27.34
5. Others						
Dry Vegetation	218.11	191.03	142.73	111.77	624.18	542.8
Open Sand	45.34	12.38	29.66	49.32	3.02	82.62
Total	3020.31	3303.33	3320.42	3315.82	3313.97	3313.78

Antwi *et al.*

Antwi *et al. SpringerPlus* 2014 3:702, doi:10.1186/2193-1801-3-702

Forest lands in the post-mining landscape include afforestation (made up of pine and deciduous tree), deciduous tree afforestation and afforestation of pine trees. Except afforestation which occupied 25.47 ha in 1988, none of the forest LC types existed in 1988 (Figures 4, 5 and Table 2). By 2003, there was no afforestation land. On the other hand; deciduous tree afforestation generally increased from 12.67 ha in 1991 to 34.47 ha and 107.89 ha in 1995 and 1998 respectively. However, by 2000, deciduous afforestation decreased to 97.77 ha and further decreased to 6.27 ha in 2003. Since its introduction in 1991, afforestation of pine trees was the only forest land that consistently increased in all the study years. From 108.14 ha in 1991, afforestation of pine trees experienced consistent increase in area and recorded 467.54 ha in 2003. The general observation was that, while afforestation and deciduous tree afforestation were decreasing, afforestation of pine trees was increasing.

Grasslands LC distributions in Schlabendorf Süd generally involved dry grassland, mixed grassland and trees and sparse pioneer grassland. There was a general decrease in all the grasslands' LCTs from 1988 to 2003 (Figures 4, 5 and Table 2). In 1988, dry grassland covered 1299.55 ha but this decreased to 1084.71 ha in 1991, though it increased to 1222.43 ha in 1995. However, in 1998, there was a decrease to 1154.39 ha. By 2000 dry grassland significantly decreased to 776.28 ha and again in 2003 to 505.81 ha. Mixed grassland and trees decreased from 114.63 ha in 1988 to 671.30 ha in 2003. The biggest decrease in mixed grassland occurred in 2000 when it decreased to 89.89 ha (Figures 4, 5 and Table 2).

Water bodies involved lake and wetland LCTs. While lake coverage generally increased throughout the study period, wetlands decreased (Figures 4, 5 and Table 2). At the start of the study in 1988, lake covered 53.6 ha, but by the end of the study year in 2003, lake had increased more than four times its area to 232.18 ha. The year 2000 recorded the lowest decrease in wetlands size, recording 0.95 ha. In 1988, wetlands occupied 72.99 ha, this generally decreased to 27.34 ha in 2003. Wetlands in Schlabendorf Süd showed a more unsettling distribution.

The open and dry vegetation LCTs consist of dry vegetation and open sand. In 1988, dry vegetation occupied 218.11 ha, but this decreased consistently to 191.03 ha, 142.73 ha and 111.77 ha in 1991, 1995 and 1998 respectively. In the case of open sand, between 1988 and 2003, there was an increase from 45.34 ha to 82.62 ha. It is however important to emphasize that, the years 1991 and 2000 recorded the lowest declines in open sand LCT of 12.38 ha and 3.02 ha respectively. Open sand areas are created as results of failed vegetation growth in afforested areas and locations of harvested pine trees. Others open sand areas are the results of dunes (Figures 4, 5 and Table 2).

Land Cover Distribution in Schlabendorf Nord from 1988 to 2003

In Schlabendorf Nord agricultural land generally decreased in size between 1988 and 2003, from 1038.58 ha to 872.30 ha respectively (Figures 5 and Table 3). Though there was an increase in agricultural land in 1991 from the previous study year, the decrease started in 1995 and continued to 1998, recording just about 856 ha. However, in 2000, agricultural land increased to 935.47 ha, but this was not sustained as there was a further decrease in 2003 to 872.30 ha.

Table 3: Land covers distribution in Schlabendorf Nord from 1988 to 2003

Land use division (Nord)	1988	1991	1995	1998	2000	2003
1. Agriculture						
Agricultural Land	1038.58	1065.84	1050.05	856	935.47	872.3
2. Forest Land						
Afforestation	302.98	16.79	114.54	12.33	34.12	16.74
Deciduous Trees Afforestation	11.48	225.11	180.25	306.11	199.11	215.99
Afforestation of Pine Trees	200.03	334.2	372.74	363.15	386.55	364.18
3. Grasslands						
Dry Grassland	146.74	60.71	141.66	272.23	280.8	153.54
Mixed Grassland and Trees	221.16	238.56	182.94	178.17	69.34	349.17
Sparse Pioneer Grassland	34.05	80.12	45.02	78.37	135.9	17.56
4. Water Bodies						
Lake	183.11	154.29	139.73	249.61	285.33	298.96
Wetland	0	0	11.89	0	0	11.16
5. Others						
Dry Vegetation	244.64	189.64	100.58	40.1	63.28	44.64
Open Sand	6.91	24.65	57.28	36.7	1.35	46.71
Total	2389.68	2389.91	2396.68	2392.77	2391.25	2390.95

Antwi *et al.*

Antwi *et al. SpringerPlus* 2014 3:702, doi:10.1186/2193-1801-3-702

Forest LC distributions involved afforestation, deciduous tree afforestation and afforestation of pine trees. Except afforestation which decreased in size between the study periods, deciduous tree afforestation and afforestation of pine trees generally increased in size from 1988 to 2003 (Figure 5 and Table 3). In 1988, afforestation land occupied 302.98 ha. However, by 2003, this had decreased to 16.74 ha. Over the period, the highest decline in afforestation occurred in 1998 where it occupied just 12.33 ha.

Deciduous forest recorded the highest increase in size from 11.48 ha to 215.99 ha in 1988 and 2003 respectively (Figure 5 and Table 3). However, the increase in year was inconsistent. From a low base of 11.48 ha in 1988, deciduous tree afforestation increased to 225.11 ha in 1991, fell to 180.25 ha in 1995, but increased again to 306.11 ha in 1998 before decreasing again in 2000 and finally picking up to 215.99 ha in 2003. Afforestation of pine trees recorded a consistent increase in size from 1988, 1991 and 1995, i.e., 200.03 ha, 334.20 ha, and 372.74 ha respectively. In 1998 afforestation of pine trees recorded a decrease in size to 363.15 ha from the previous study year, but this increased slightly to 386.55 ha before declining marginally again to 364.18 ha in 2003.

While both dry grassland and mixed grassland and tree generally increased in size from 1988 to 2003, sparse pioneer grassland generally decreased over the same period. It is important to emphasize that, the increase recorded in both dry grassland and mixed grassland and tree were not consistent throughout the study period. For instance, in 1988, while dry grassland recorded 146.74 ha, by 1991, this decreased to 60.71 ha, though there was an increase to 141.66 ha by 1995, 272.23 ha in 1998 and 280.80 ha in 2000. In 2003, dry grassland recorded 153.54 ha, which is a decrease from the previous study year. More sparse pioneer grassland was introduced in Schlabendorf Nord in 1991 and 1998 (Figure 5 and Table 3).

Lakes and wetlands formed the main water bodies in Schlabendorf Nord. While Lake LCT was present in all the study periods, wetlands only emerged in 1995 and 2003 (Figures 4, 5 and Table 3). In 1995 the wetland area was 11.89 ha. However, by 2003 when it reappeared on the LC scene this had decreased marginally to 11.16 ha. It has to be emphasized that the location of the 11.16 ha wetland in 2003 was different from that of 1995. On the other hand, the lake generally increased from 183.11 ha in 1988 to 298.96 ha in 2003. Apart from 1991 and 1995 when the lake seize decreased, there was a consistent increase in lake size in 1998, 2000 and 2003.

Open sand and dry vegetation constitute vegetation free areas and dry vegetation LC distributions in Schlabendorf Nord. While dry vegetation decreased in size from 1988 to 2003, open sand increased in size within the same period, though in both cases the changes were not consistent (Figures 4, 5 and Table 3). From 244.64 ha in 1988, dry vegetation decreased to 189.64 ha in 1991 and to 100.58 ha in 1995. The most significant decrease occurred in 1998 (40.10 ha), though by 2000 there was a marginal increase to 63.23 ha and a final decline to 44.64 ha in 2003. Open sand on the other hand increased from 6.91 ha to 46.71 ha from 1988 to 2003. However, the year 2000 recorded a significant decline in open sand area of 1.35 ha.

Implication of LCCs on Vegetation Succession and Water Bodies in the Post-Mining Landscape

Land Cover Change and Primary Succession in Schlabendorf Post-Mining Landscape

LCC involving primary succession generally had three main LCTs: agricultural land, dry grassland and pioneer grassland changing to different LCTs. These are discussed in three segments. Segment one spans between 1988 and 1991, segment two is between 1995–1988 and segment three from 2000 to 2003. In Schlabendorf Nord, from 1995 to 1998, agricultural lands were abandoned for either dry grassland (14.28 ha) or mixed grassland and trees (5.46 ha). On the other hand, dry grassland mainly changed to dry vegetation and grassland and trees (Table 4). While changes to dry vegetation occurred in both Schlabendorf Süd and Nord in the first segment between 1988 and 1991, changes to grassland with trees first occurred only in Schlabendorf Süd in the first segment and then in both Schlabendorf Süd and Nord in the third segment (Table 4).

Table 4: Land cover changes in the post-mining landscape at different study periods

Type of land use changes	1988 - 1991		1995 - 1998		2000 - 2003	
	Nord	Süd	Nord	Süd	Nord	Süd
Primary succession						
Agricultural Land to Dry Grassland			14.28			

Agricultural Land to Grassland and Trees			5.46			
Dry Grassland to Dry Vegetation	22.93	6.74				
Dry Grassland to Grassland and Trees		5.99			31.01	24.37
Pioneer Grassland - Grassland and Trees		4.29			21.93	7.23
Pioneer Grassland to Dry Vegetation						16.33
Pioneer Grassland to Dry Grassland			53.54		52.33	31.62
Open Sand to Pioneer Grassland		78.13				
Wetland to Pioneer Grassland		25.69				
Dry Vegetation to Grassland and Trees	34.60				12.33	
Lake to Grassland and Trees	4.59					
Afforestation						
Agricultural Land - Pine Afforestation				11.72		
Agricultural Land - Afforestation		6.51				
Agriculture Land to Deciduous Afforestation	4.55		5.04		10.13	
Dry Grassland to Afforestation		3.54				5.19
Dry Vegetation to Afforestation		12.49				
Afforestation to Pine Afforestation	23.81	97.05	15.69			49.68
Afforestation to Deciduous Afforestation	52.90		42.86			

Deciduous Afforestation to Pine Afforestation						30.18
Grassland and Trees to Pine Afforestation	26.78		5.79	21.57		15.79
Grassland and Trees to Deciduous Afforestation				6.40		
Reversed succession						
Afforestation to Grassland and Trees					72.65	38.36
Deciduous Afforestation to Grassland and Trees					28.42	34.76
Dry Vegetation to Dry Grassland	37.37	29.00		37.33		
Dry Vegetation to Open Sand						2.53
Dry Vegetation to Agriculture Land	12.71				37.19	
Dry Grassland to Open Sand		0.67		2.23	6.21	3.83
Dry Grassland to Agriculture Land		3.16			32.04	
Pioneer Grassland to Dry Vegetation		3.52				
Pioneer Grassland to Open Sand				1.46		3.26
Afforestation to Agriculture Land	11.21					
Pine Afforestation to Grassland and Trees					6.58	2.91
Dry Grassland to Pioneer Grassland	34.13		17.38			
Grassland and Trees to Dry Grassland				19.45		

Grassland and Trees to Agriculture			17.79			
Grassland and Trees to Agriculture Land	22.34					
Lake formation						
Dry Grassland to Lake		0.64	25.80	1.73	4.63	3.28
Dry Vegetation to Lake		6.18		2.23	7.5	5.07
Open sand to Lake		6.26	45.82	38.14		40.93
Pioneer Grassland to Lake		1.25		0.43	4.89	2.76
Wetland to Lake		7.80	88.25	35.27		65.61
Grassland and Trees to Lake			10.01		5.78	
Dry Grassland to Wetland				0.69		
Pioneer Grassland to Wetland				0.57		1.40
Agriculture Land to Wetland					1.20	

Antwi *et al.*

Antwi *et al. SpringerPlus* 2014 3:702, doi:10.1186/2193-1801-3-702

In Schlabendorf, pioneer grassland, made up of mostly diaspora hay mulch, was mainly used as primary colonizer. Typically, this was seeded grassland. During the early days of succession, pioneer grassland, which was introduced to often bare areas, was replaced by grassland with trees, dry vegetation and dry grassland. The changes to grassland and trees occurred first in Schlabendorf Süd in the first segment (Figure 6). In the first segment, 4.29 ha of pioneer grassland was replaced by mixed grassland and trees, while in the last segment the replaced area came to 21.93 ha in Nord and 7.23 ha in Süd. The replacement of pioneer grassland with dry grassland on the other hand occurred only in Schlabendorf Süd in the last segment. Land cover changes from pioneer grassland to dry grassland took place in Schlabendorf Nord in both the second and third segments and in Schlabendorf Süd in the third segment only (Table 4).

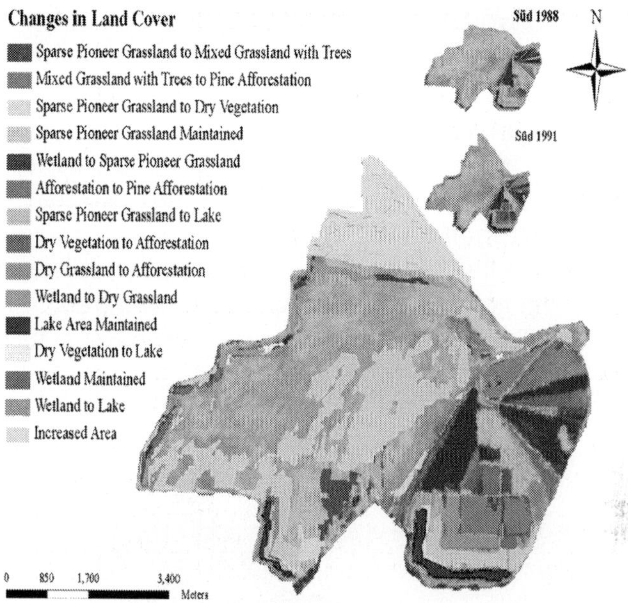

Figure 6: LCC illustrations at Schlabendorf Süd from 1988 to 1991.

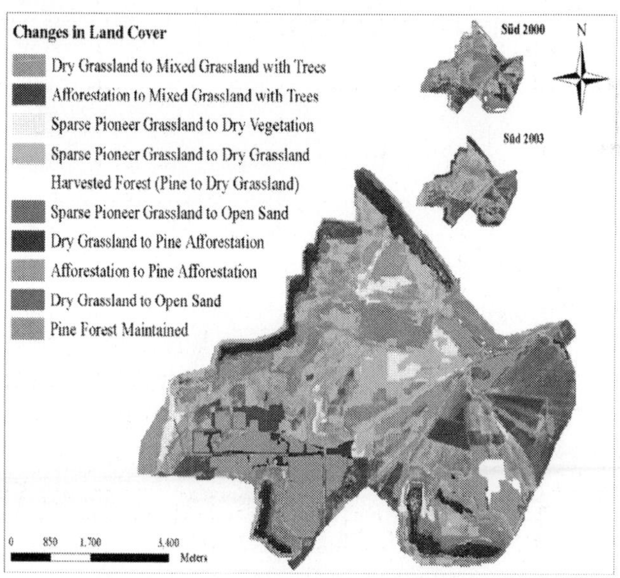

Figure 7: LCC illustrations at Schlabendorf Süd from 2000 to 2003.

Besides the above successional changes, other LCCs contributed to the primary succession stage of the post-mining landscape. These include planting 78.13 ha of sparse pioneer grassland in open sand areas in Schlabendorf Süd from 1988 to 1991. During the same period in the same landscape, 25.69 ha of wetlands area was replaced with sparse pioneer grassland (Figure 6). Other successional changes in Schlabendorf Nord include significant tree growth in the grassland (during the first and last segments) which saw the replacement of dry vegetation LCT replaced with mixed grassland and tree LCT. A portion of lake (4.59 ha) changed to mixed grassland and trees during the first segment (Table 4).

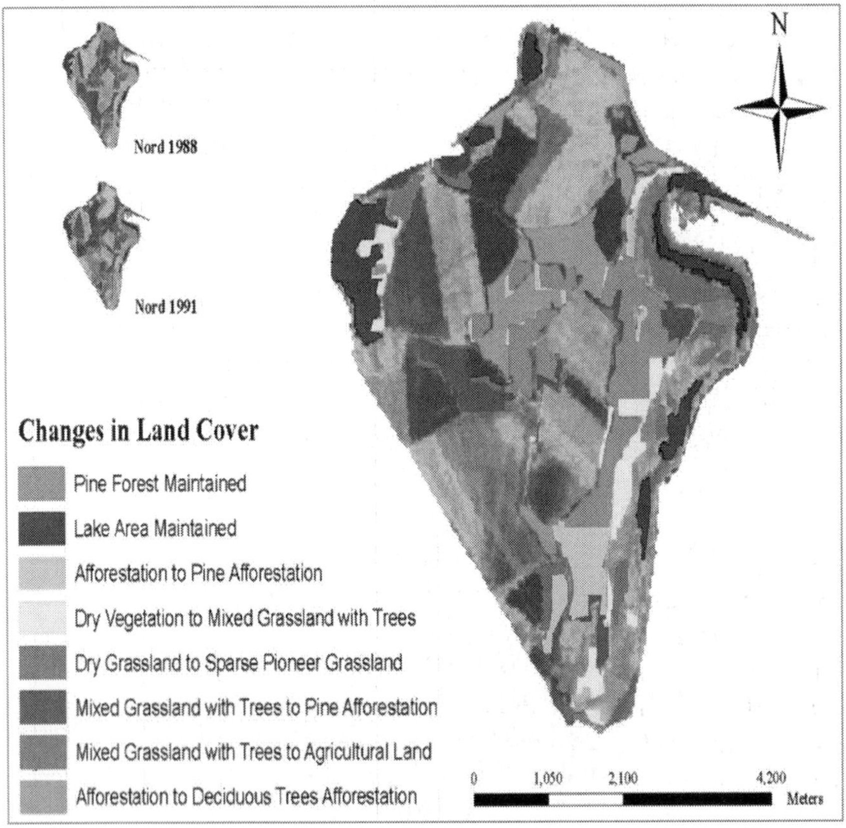

Figure 8: LCC illustrations at Schlabendorf Nord from 1988 to 1991.

Land Cover Change and Post-Mining Landscape Afforestation

Most of the arable lands were abandoned for afforestation. Agricultural lands were generally replaced by afforestation of pine trees, deciduous afforestation and afforestation. In Schlabendorf Süd, afforestation from agricultural land occurred in the first segment (Table 4). However, replacing agricultural land with deciduous tree afforestation only occurred in Schlabendorf Nord in all the three segments. Apart from agricultural lands, other LCTs that contributed to forest growth are mixed grassland and trees and afforestation (Figures 6, 7, 8 and Table 4). Mixed grassland and trees contributed to the formation of pine and deciduous afforestation. The gains in afforestation of pine trees occurred in both landscapes. In Schlabendorf Nord, the gains occurred in the first segment and second segment with 523 26.78 ha and 5.79 ha respectively. In Schlabendorf Süd 21.57 ha and 15.79 ha gain occurred in the second and third segment respectively. In Schlabendorf Süd only 6.40 ha of mixed grassland and trees was lost to deciduous afforestation (Table 4).

Dry grassland and dry vegetation were both used for forest growth in the afforestation LCT. Changes from dry grassland to afforestation occurred only in Schlabendorf Süd where a change of 3.54 ha and 5.19 ha occurred in the first and third segments respectively. On the other hand, dry vegetation LCT was replaced with afforestation only in Schlabendorf Süd during the first segment, with 12.49 ha of dry grassland involved (Table 4).

Among the afforested lands, afforestation of pine trees and deciduous afforestation gained more from the losses in other LCTs. However, there were instances where deciduous afforestation also lost to afforestation of pine trees (Table 4). While other LUTs were changing to afforestation (mixed forest), afforestation often lost to afforestation of pine trees and deciduous tree afforestation. In Schlabendorf Nord, 23.81 ha and 15.69 ha of afforestation was replaced by afforestation of pine trees in the first segment and second segments. In Schlabendorf Süd, however, replacement of afforestation by afforestation of pine trees occurred in the first and last segments. In the first segment 97.05 ha of afforestation was replaced by afforestation of pine trees (Figures 6 and 8) in the last segment the coverage was 49.68 ha (Table 4).

Land Cover Change and Vegetation Reversed Succession

Reverse succession occurred mostly from changes in dry vegetation to either dry grassland, open sand, or agriculture land, while dry grassland also changed to open sand, agriculture land and pioneer grassland. Changes from dry vegetation to dry grassland mostly occurred in Schlabendorf Süd in the first and second segments from 29 ha to 37.33 ha respectively. However, in the first segment in Schlabendorf Nord 37.37 ha of dry vegetation changed to dry grassland. On the other hand, changes from dry vegetation to agricultural land occurred only in Schlabendorf Nord in the first and last segment from 12.71 ha to 37.19 ha respectively. LLCs from dry vegetation to open sand occurred only in the last segment in Schlabendorf Süd where 2.53 ha were involved (Table 4).

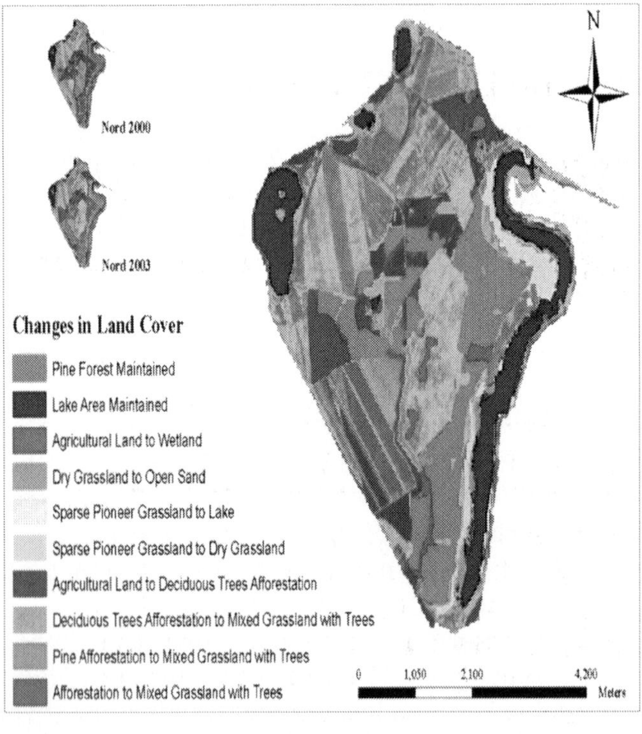

Figure 9: LCC illustrations at Schlabendorf Nord from 2000 to 2003.

Dry grassland was lost to open sand, pioneer grassland and agriculture land (Figures 8, 9 and Table 4). Changes in dry grassland to open sand mostly occurred in Schlabendorf Süd in the first, second and third segments from 0.67 ha, 2.23 ha and 3.83 ha respectively. In the third segment, 6.21 ha of dry grassland changed to open sand in Schlabendorf Nord. In the first and second segments in Schlabendorf Nord, 34.13 ha and 17.38 ha of dry grassland changed to pioneer grassland (Figure 8 and Table 4). Pioneer grassland was also lost to dry vegetation and open sand.

Changes from pioneer grassland to open sand occurred in only Schlabendorf Süd in the second and third segments (Figure 7 and Table 4). Other notable reverse successions involved changes from afforestation to agriculture land and from afforestation of pine trees to mixed grassland and trees. 11.21 ha of afforestation land changed to agriculture land only in the first segment, while 6.58 ha and 2.91 ha of afforestation of pine trees changed to grassland and trees respectively (Figure 7 and Table 4).

Land Cover Change and Formation of Lake

The major LCTs that were converted to lake include dry grassland, pioneer grassland, dry vegetation, open sand, wetland, grassland and tree and agricultural land. In 1988, there was no wetlands in Schlabendorf Nord (Figures 6, 9, 10, 11 and Table 3). Conversion of Agricultural land to wetlands recorded the least change with just 1.20 ha involved.

Estimation of Primary Production from Normalized Differential Vegetation Index

NDVI values in Schlabendorf Nord showed greater variation in primary production from 1988 to 2003. The highest NDVI values in Schlabendorf Nord and Süd were in 1991, 1998 and 2003 (Figure 10). Unlike Schlabendorf Süd which had a largely consistent increase in NDVI values, Schlabendorf Nord had fluctuating and generally greater NDVI values. In both landscapes, the least primary production was observed in 2000. Schlabendorf Süd had a consistent increase in primary production except in 2000. In Schlabendorf Süd, though primary production decreased in 2000 to its initial value in 1988, there was a sharp increase in 2003 when the highest NDVI value was recorded. The years 1988 and 2000 recorded the least primary production.

DISCUSSIONS

Land Cover Distributions and Change Analysis in Schlabendorf Nord and Schlabendorf Süd

Land Cover Distributions and Change Analysis in Schlabendorf Nord

In Schlabendorf Nord, vegetation growth was mainly in the form of pioneer grassland and dry grassland changing to dry vegetation which, together with agricultural land, were changed to mixed grassland and trees. Subsequently, all the forest LCTs (afforestation, deciduous afforestation and afforestation of pine trees) were derived from mixed grassland and trees. In Figure 11 successional pathway in Schlabendorf Nord is shown. In terms of succession successes, pine forest and mixed grassland represent significant restoration success (in areas intended for recultivation) and contributed greatly to the increasing net primary production in most periods of assessment. Agricultural land also played a significant role in the forest conversion witnessed in Schlabendorf Nord, as most of the agricultural lands were abandoned for forest growth (Figure 11and Table 4). Bungart and Hüttl (2001) found that the decline in agricultural land enabled forest recovery in Schlabendorf. Sustainable LU and an increase in forest area could occur if forest conservation, agricultural abandonment, and forest-cutting cycles are managed specifically to increase forest extent (Drummond and Loveland 2010). Nevertheless this assertion strongly depends on intended reclamation goal. Interestingly, agricultural land emerged as the most stable LCT in Schlabendorf Nord. For instance, in 1988, 1991 and 1995 (Figure 4 and Table 4), agricultural land did not undergo any significant changes. Thus, though much of its area changed to other LCTs, it also received losses from other LCTs. This also confirms the pattern that emerged in the ordination plane as agricultural land was the most centrally placed (Figure 13).

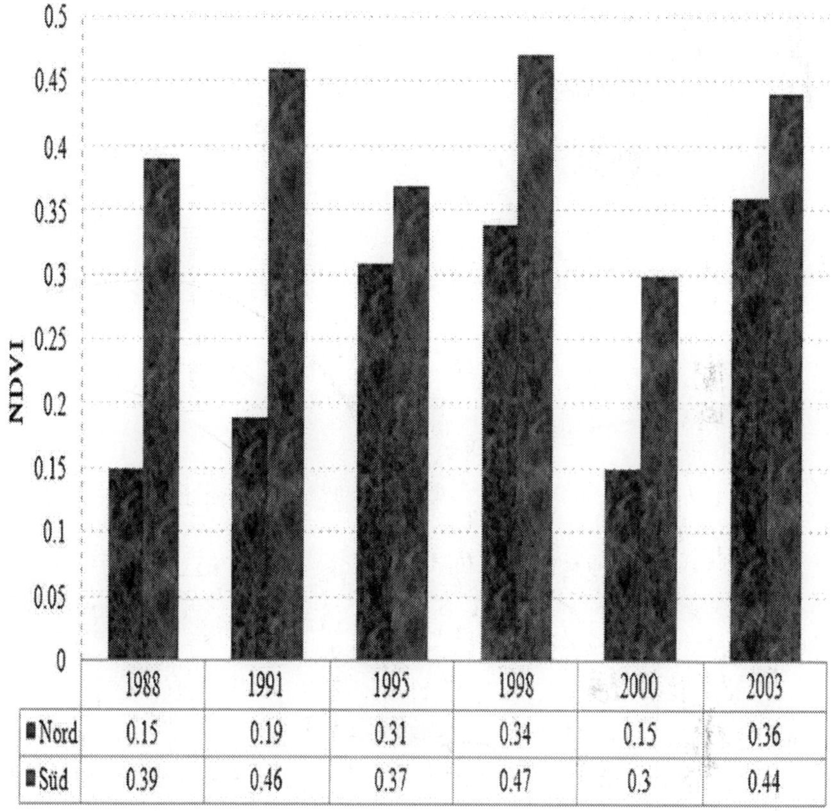

Figure 10: Primary productions for Schlabendorf Nord and Schlabendorf Süd in 1988, 1991, 1995, 1998, 2000 and 2003.

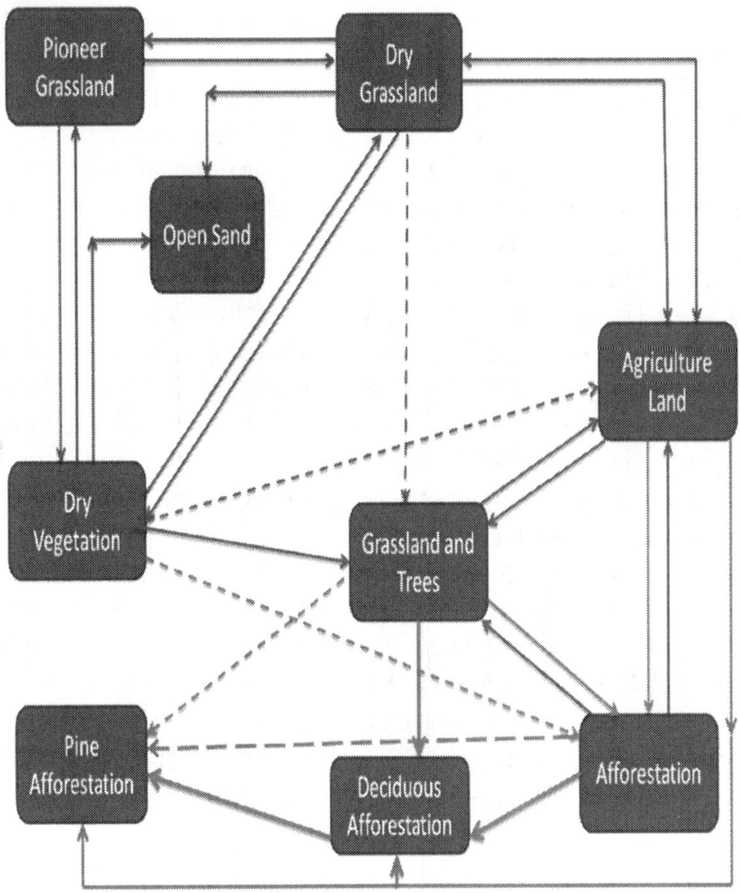

Figure 11: Land cover changes and successional pathways in Schlabendorf Nord.

In Schlabendorf Nord, there was barely any recorded transition between open sand and pioneer grassland (Figure 11). This suggests that the vegetation growth in the old post-mining landscape has passed the level where primary colonizers grow from bare land and sand dunes. On the other hand, some vegetated areas, e.g., dry grassland and dry vegetation, became open sand, giving an indication of reverse succession. It was notable that agricultural land contributed significantly to reverse succession as LCTs such as afforestation, mixed grassland and trees, dry vegetation and dry grassland areas were turned into agricultural land (Figure 11 and Table 4).

Most of the afforestation land was lost to deciduous afforestation and afforestation of pine trees. This is visible from the continuously increasing size of afforestation of deciduous trees and decreasing size of afforestation. Any increase in deciduous afforestation and afforestation of pine trees LC corresponded to decreased afforestation LCT in the same year of study. This dissimilarity as well existed between afforestation of deciduous trees and dry vegetation. This explains why they are positioned at the extreme opposite ends of the ordination plane (Figure 13). The dissimilarity in distribution pattern of afforestation and deciduous tree afforestation LCTs is due to the fact that afforestation was made up of a mosaic of free growing pine and deciduous trees (mixed forest), which was gradually cultivated predominantly with deciduous trees and partly with pine trees (Table 4).

The pattern of distribution of afforestation was similar to that of dry vegetation in most of the period of study (Figures 4, 13 and Table 3). But since afforestation generally decreased at the end of 2003, increase in the dry vegetation in this case supports reverse succession. This inference is made from the general increase in dry vegetation area but not strictly linked to decreased size of afforestation.

Some LC distribution patterns were similar or close to, but not dependent on others. An example in Schlabendorf Nord is where mixed grassland and open sand largely had similar LC distribution for the most part of the study except the variation in 1991.

Land Cover Distributions and Change Analysis in Schlabendorf Süd

In Schlabendorf Süd, the transition in LCTs between the lower vegetation cover (e.g., pioneer grassland, dry vegetation, and dry grassland) and forest land is mixed grassland and trees. The succession pathways towards afforestation involve conversions of pioneer grassland and dry grassland to grassland and tress; mixed grassland and trees were then converted to afforestation, afforestation of pine trees or deciduous forest. In addition, dry vegetation and agricultural land also contributed to forest land gains (Figure 12 and Table 4). The successional pathway of Schlabendorf Süd is shown in (Figure 12). LC distribution patterns involving afforestation of pine trees, lake and mixed grassland in Schlabendorf Süd were mostly similar, though this similarity does not indicate any ecological interdependence or process. They lie close to each other on the same side of the ordination plane because they generally increased in size during the study period (Figure 13 and Table 4). Pine forest and mixed grassland represent significant restoration success (in areas intended for forest growth) and contributed greatly to the increasing net primary production in most periods of assessment (Figure 12).

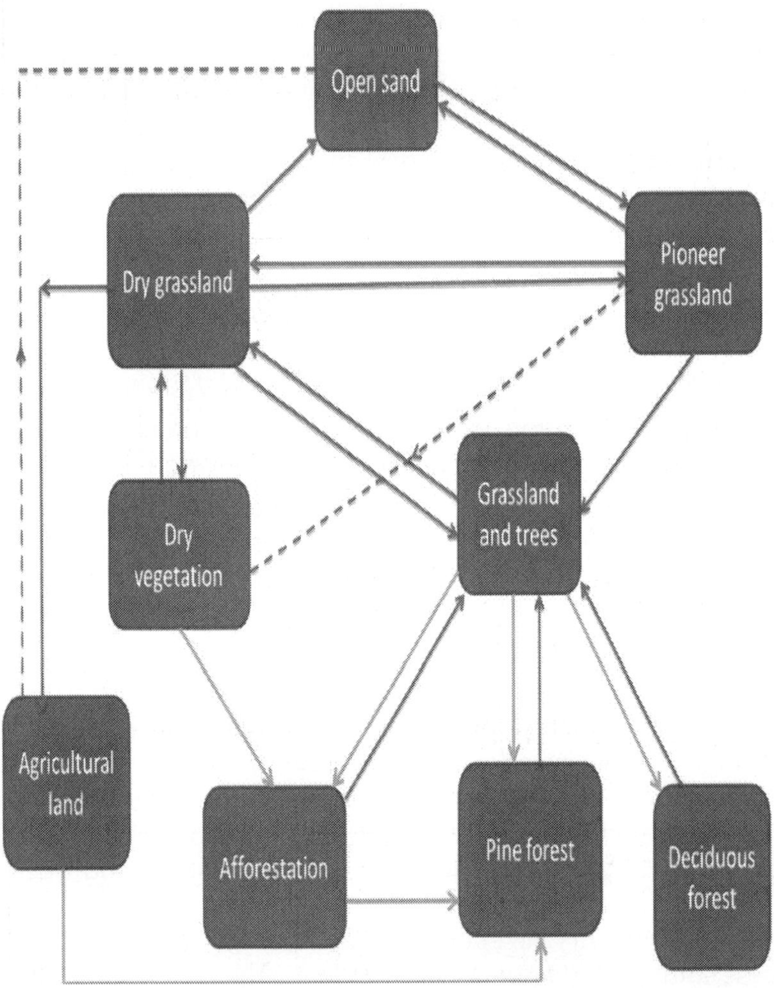

Figure 12: Land cover changes and successional pathways in Schlabendorf Süd.

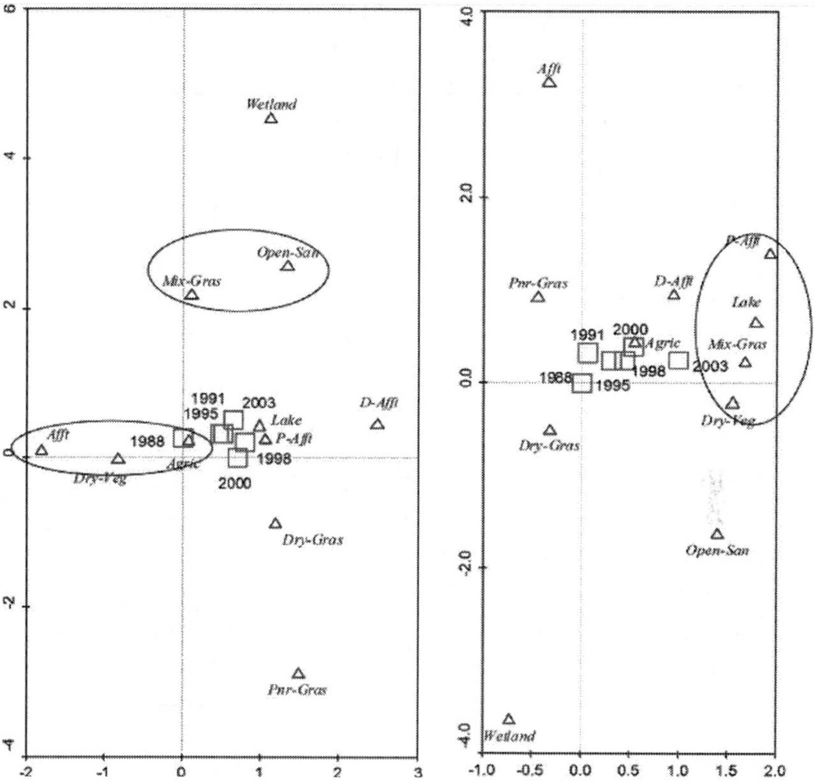

Figure 13: DCA ordination plot for land cover distribution at Schlabendorf Nord (Left) and Schlabendorf Süd (Right) with untransformed data. Legend: P-Afft - represents afforestation of pine trees, Afft - afforestation, D-Afft - deciduous trees afforestation, Dry-Veg - dry vegetation, Agric - agriculture land, Mix-Grass - mixed grassland and trees, Pnr-Grass - sparse pioneer grassland, Dry-Grass - dry grassland, Open-Sand - open sand, Wetland - wetland, Lake - lake.

While in the case of Schlabendorf Nord, deciduous forest was changed to both afforestation and/or pine forest, in the case of the Schlabendorf Süd, deciduous forest rather lost to grassland with trees, indicating a reverse succession. In addition to this, both pine forest and afforestation also lost to mixed grassland with trees. Other notable forms of reverse succession involved changes from agricultural land, dry grassland and pioneer grassland to open sand (Figure 12 and Table 4). For instance, dry grassland cover experienced a reverse of the parallel transformation observed with regard to afforestation of pine trees, lake and mixed grassland throughout the study period. This is a typical case where reverse succession occurred in the landscape at Schlabendorf Süd. These transformations show that in Schlabendorf Süd there were concurrent back and forth gains and losses in LCTs. These transformations suggest succession and reverse succession among LCTin most cases (Felinks 2000). This is confirmed by the fact that, for most periods studied in Schlabendorf Süd, LC distribution pattern of agricultural land and deciduous tree afforestation were similar, though some variation occurred in both vegetation types in 1998 and 2003 (Figure 13 and Table 4). An increase in one corresponded to an increase in the other (Table 2). Here portions of deciduous trees in a naturally growing area with pine trees were replaced with afforestation of pine trees and others also emerged as mixed grassland with trees in 2003. Agricultural land did not change significantly.

As exemplified in the ordination plane, most LCTs in Schlabendorf Süd (4 out of the 11) were located on the left side of the first ordination axis as compared to 2 out of 11 LCTs in Schlabendorf Nord (Figure 13). The location of afforestation, pioneer grassland and dry grassland at Schlabendorf Süd implies decreasing LC area over the study period that favours reverse succession. The greater part of the sown grassland failed. Change detection results reveal that a greater part of the area that was lost in pioneer grassland became open sand (Table 4). Thus reverse succession occurred much more in Schlabendorf Süd as compared to Schlabendorf Nord.

Comparative Land Cover Distribution and Change in the Two Post-Mining Landscapes

While in Schlabendorf Nord afforestation was replaced by afforestation of pine trees and deciduous afforestation, in Schlabendorf Süd afforestation was replaced by only afforestation of pine trees (as forest land) and mixed grassland and trees. In Schlabendorf Süd, it was more common for forest to be lost to grassland, but in the Schlabendorf Nord this was less common. In Schlabendorf Süd, all the forest lands were lost to grassland, but in Schlabendorf Nord, only afforestation was lost to grassland, suggesting more reverse succession in Schlabendorf Süd (Figures 11 and 13). Until recently, in Schlabendorf, seeded grassland were used as primary colonizers to reclaim the mine-damaged land. On the other hand, reverse succession occurred in both landscapes.

An interesting trend of reverse succession was the frequent change of dry vegetation/dry grassland/sparse pioneer grassland to open sand, open sand to wetland; and wetland to lake (Figures 11 and 12). Felinks (2000) described plant succession in a post-mining landscape as having stages of stochastic regression instead of web-like stages of succession. Reverse succession in this context is defined as change from a particular LCT after a period of vegetation growth back to the initial or a characteristically comparable LCT via natural or artificial processes. Another notable difference in LC distribution between Schlabendorf Nord and Schlabendorf Süd was that, while in the North there was no recorded transition between open sand and pioneer grassland, in the Süd this transition was clearly visible. This suggests that vegetation growth in the Nord was not from the primary level as was the case with the Süd.Also, while in Schlabendorf Nord, agriculture land played an essential role in forest growth and emerged as the most stable LCT because it was the most centrally located LCT in the ordination plane; in Schlabendorf Süd agricultural land was barely involved in forest growth, though it was the most stable LCT (Figures 11 and 13).

Unlike Schlabendorf Nord, the LC distributions in Schlabendorf Süd corresponded with the gradient of time.

Effect of Land Cover Change on Formation of Lake in Schlabendorf Post-Mining Landscape

In Figures 14 and 15, areas of the post-mining landscape occupied by lake and wetlands are shown. Greater areas of water body in both post-mining landscapes were maintained (Figures 6and 9). In Schlabendorf Süd there was constant increase in lake area during the study except in 1995. Since 1988, lake area has increased by four times its size (53.60 ha), with the most significant gain occurring between 2000 and 2003 (Figures 4, 5 and 14 and Table 2). Increase in lake size contributed to the reduced wetland area. However, only the first segment had a reduction in wetland area. A similar situation occurred in dry grassland and sparse pioneer grassland which also had a direct effect on reduced area of wetland. The ongoing transformation in lake and wetlands are mainly due to differences in age since dumping of spoils in the post-mining area, increased LU intensity and difference in water table levels in both landscapes. In Schlabendorf Nord, out of the 555.30 ha of land expected to become lake by 2010, about 300 ha had already formed lake as of 2003. However, only about 232 ha of the 659 ha of land expected to form lake by 2010 in Schlabendorf Süd had become lake in 2003 (Der Braunkohlenausschuss 1994). This can be explained by the fact that the old post-mining landscape (Schlabendorf Nord) has attained greater water table balance while the young post-mining landscape still needs some time to attain this balance. According to Grünewald (2001), the 'greatest number of lakes in the Lusatia region will be filled in about 25 years' time'. This could lead to fragmentation due to relocation of reclamation related roads and other construction activities. Following Grünewald's finding and our observations, parts of dry vegetation, sparse pioneer grassland, dry grassland, open sand and wetland that are already changing to "lake" or share a direct boundary with "lake" would likely be affected by the increasing size of "lake". The vegetation in these LUTs in both landscapes has a shallow fibrous root system or open sand which cannot serve as Lake Bank protection.

In order to operate an open cast mine, groundwater level should be pumped to beneath the deepest working level (Grünewald 2001). Since the mine closure in 1991 in Schlabendorf, pumping rate dropped, so that from 31.80 m^3/s it reduced to 17 m^3/s in 2000, leading to a rising groundwater table (Grünewald 2001), hence the significant increase in lake size in both landscapes.

The decreased lake size in 1995 could be due to changes in precipitation and high surface temperature. In August 1995 (satellite imagery acquisition date), the monthly average precipitation was below the mean value from 1961 to 1990. Furthermore, July 1995 was the hottest summer in the 15 years studied. Thus the reduced lake and wetlands area could be attributed to high surface temperature and dry summer conditions (Deutscher Wetterdienst).Change in wetlands revealed a remarkable difference in LCC in both landscapes. Figure 15 shows wetland was mostly absent in Schlabendorf Nord. On the contrary, Schlabendorf Süd always had large areas of wetlands, with some measuring six times the largest wetland in Schlabendorf Nord (Figures 14 and15). Thus the young post-mining landscape has an unstable groundwater table compared to the rather stable older landscape. This is the reason for the constant occurrence of wetlands in Schlabendorf Süd which in most cases turn into lake (Figure 14).

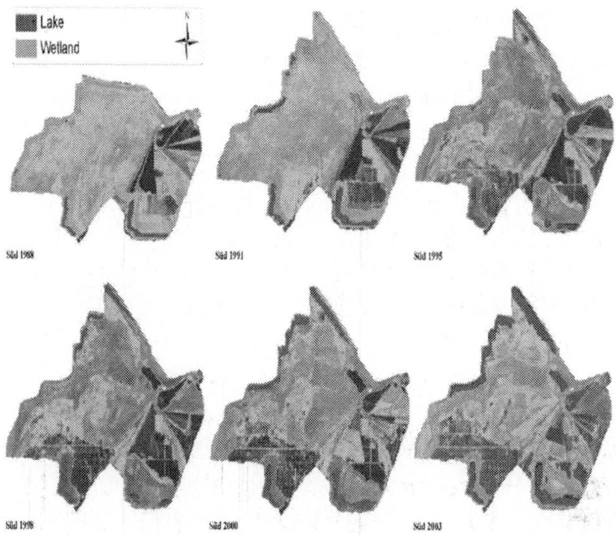

Figure 14: Primary productions for Schlabendorf Nord and Schlabendorf Süd in 1988, 1991, 1995, 1998, 2000 and 2003.

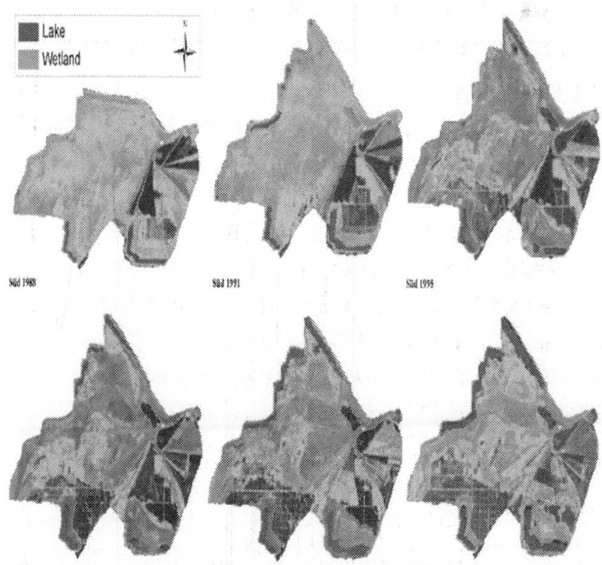

Figure 15: Area of land occupied by lake in Schlabendorf Nord during 1988, 1991, 1995, 1998, 2000 and 2003.

Biomass Accumulation (Primary Production) in Schlabendorf

The afforested areas in the young landscape experienced a steady increase in primary production (about 25% pine and deciduous trees plantation) as intended in the reclamation goal (Bungart and Hüttl 2001). Furthermore, most of the afforestation of pine trees was maintained or increased continuously. This is reflected in the 44% biomass accumulation (Figure 12). Significant increase in mixed grassland also contributed greatly to succession successes and overall increase in biomass accumulation.

The fluctuating net biomass accumulated resulted from changing areas of dry grassland, dry vegetation, lake and open sand. NDVI values range from -1 to +1 but vegetation values typically fall within 0 to +1 with higher (greener) index values representing more active growth and primary production.

Effect of Soil Condition on Land Cover Change in the Post-Mining Landscape

An important question that emerged was, why did some cultivated areas (e.g., afforestation of pine trees) turn into open sand? Incidentally, those failed afforested areas were along the path of a mine strip which probably had lower soil pH to support healthy plants. Soil pH in the area of failed afforestation of pine trees (along the path of the mine strip) were generally low (Figure 16), while those taken from the pine forest across the open sand areas were high on both sides of the pine forest but decreased in the open sand area. Samples taken strictly from the pine forest also had comparably high pH values.

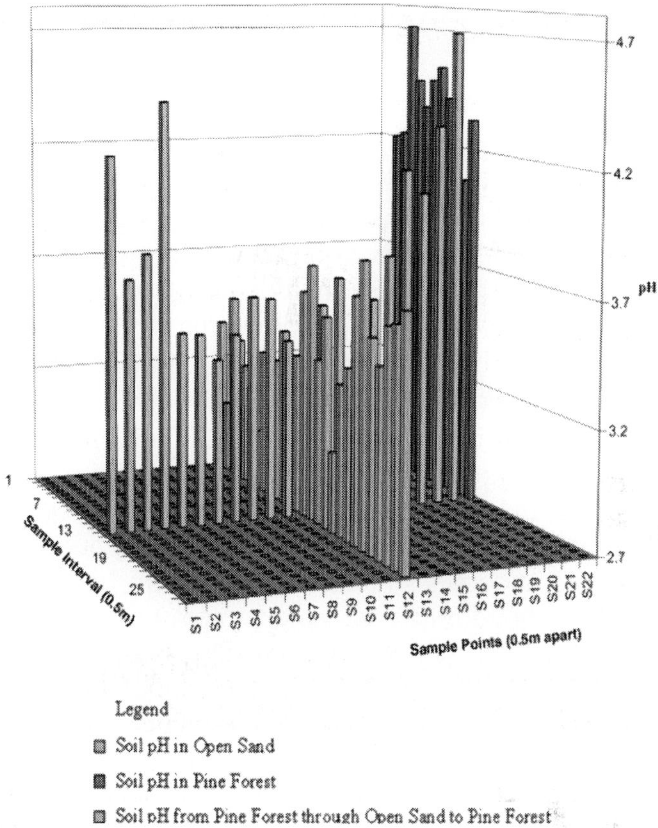

Figure 16: Soil pH taken from failed pine afforestation area and successful pine afforestation.

Whenever soil becomes acidic some nutrients may become less available for plant uptake, although plants have different nutrient needs. Soil pH (CaCl2) of 5.20 to 8 provides suitable conditions for effective plant growth (Fernández and Eichert 2009); however, most soil pH recorded was above this range. High soil acidity such as that measure in Schlabendorf Süd could reduce or weaken populations and the activity of organisms that convert N, S, and P to plant-available forms. It became evident that low pH and impoverished soil condition were some of the causes of failed vegetation growth after afforestation in parts of the landscapes. Thus low soil pH and nutrient deficient soils contributed to reverse succession in Schlabendorf (Fernández and Eichert 2009).

Reclamation of Mine-Damaged Areas in Schlabendorf

Quantification of the effects that mining activities have on ecosystems is a major issue in sustainable development and resources management (Latifovica et al. 2005). In Schlabendorf, the impacts of mining have spread beyond the site of extraction, affecting the neighbourhood of mined areas (Haigh 2000). These include constructed mining camps, roads, foot paths and contamination of nearby rivers (Spree River) by iron and soil or groundwater by acid. This situation requires some sort of restoration or intervention.

Parts of the post-mining landscape with greatly impoverished soils could be restored through some sort of human intervention (Bradshaw 1999; Hendrychová 2008; Groninger et al. 2007). However, reclamation of mine-damaged land is generally complex, often expensive and comes with specific problems based on the type of mining. In Schlabendorf, an estimated amount of 300 million Euros has gone into the reclamation program (Der Braunkohlenausschuss 1994). Bradshaw (1999) found out that reclamation schemes must focus on providing some sort of human intervention since succession may take a longer period. In Schlabendorf, such human interventions that could accelerate succession is the enrichment of the soil with nutrients before recultivation, especially in nutrient deficient areas that lost their vegetation cover. This is important since the topsoil was not replaced to re-establish soil fertility (Schultz and Wiegleb 2000). In such areas, enriching the topsoil through the application of selected diaspore enriched-heymulch could significantly facilitate primary succession. The addition of ashes (potassium) to the acidic soil is very vital to plant succession. It is worth noting that such an intervention measure is desirable in cases where the aim of reclamation is to get closed vegetation cover (e.g., towards protection from wind erosion), otherwise treatment should be restricted to areas which require them.

Causes of Land Cover Changes in Schlabendorf

In order to understand the driving forces of future trends in LUC and the prospects for sustainability in post-mining landscape restoration, we need to identify the major driving forces. Causes of LCC can be direct and indirect, though a complex interaction of the two influences LCC at varying times and in different spaces (Geist et al. 2006). Nutrient deficient soil due to a turning of the upper topsoil, prolonged seasonal fluctuations in the microclimate, low soil pH (Haigh 2000), harvesting of afforested areas (Figure 17) and other restoration related construction activities was the leading cause of the LCCs and reversed succession in both landscapes (Bröring and Wiegleb2005).

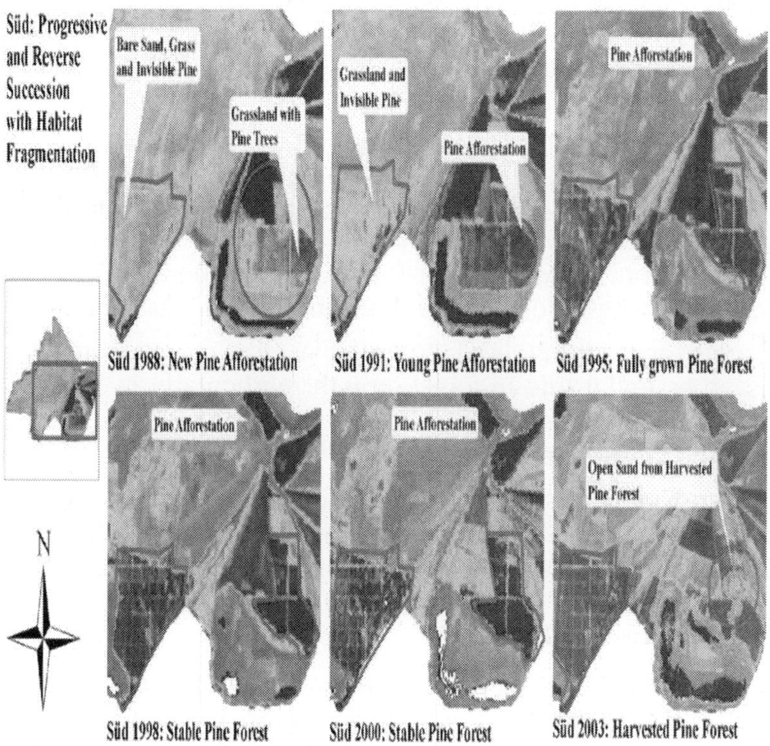

Figure 17: Progressive and reverse succession in Schlabendorf.

The process of LCC takes place at an interface between environmental and human systems interrelating with each other in a feedback mechanism (Geist et al. 2006; Demirel et al. 2011). And human policies such as *"stop pumping water from lakes"*, leveling of would-be-lake bottoms and soil compaction by explosives led to some of the observed LCC.

CONCLUSIONS

Visible patterns of increase and decrease in the LCT occurred in both landscapes. It is like a field laboratory where different habitat types interact to create new patterns of land cover types without any human societal influence such as population expansion, wealth generation etc.

Land cover transformations in the post-mining landscape showed similarity, yet it varied greatly in both landscapes. The variation reflects the impact of different ages of reclamation since dumping of spoils, land use intensity and difference in water table levels in both landscapes. Mixed grassland and tree in both landscapes was at the central point of the reclamation. Lower vegetation cover such as pioneer grassland, dry vegetation and dry grassland often change to mixed grassland and tree, which largely changed to the forest LCTs. In Schlabendorf Nord, afforestation often replaced afforestation of pine trees and deciduous tree afforestation. However, afforestation in Schlabendorf Süd was only replaced by afforestation of pine trees and mixed grassland and trees.

Though agricultural land emerged as the most stable LCT and played an essential role in forest growth, particularly in Schlabendorf Nord, agricultural land barely contributed to the afforestation of the post-mining landscape in Schlabendorf Süd.

LC transformations in the two post-mining landscapes subjected to different ages of reclamation since dumping are mostly the results of progressive and reversed plant succession. Nevertheless, the process of natural succession itself should be aided by human intervention if reclamation is to be accelerated.

Given two post-mining landscapes subjected to different ages of reclamation after dumping, clear differences in primary production and LCC pattern would occur. In Schlabendorf, the two post-mining landscapes are getting more diverse over time. The extent of these differences is partly due to the LU intensity and other geophysical factors. This is in line with the natural succession of an ecological process in an ecosystem.

During early stages of restoration, LCTs often have unstable conditions and experience more acute transformation due to increased LU intensity. This is eventually replaced by a more stable landscape with fewer transformations between the different LCTs.

Traditional mining operations like surface lignite mining as occurred within the Lower Lusatia region may not fit into the commonly understood definitions of sustainability. However, mineral extraction and later reclamation operations can contribute significantly to sustainability through the benefits they provide to society as well as the incorporation of sustainability principles in all phases of mining till its closure. This helps create a sustainable post-mining landscape suitable for predefined future LU after completion of reclamation activities and mine operation.

Land cover change takes place at the interface between environmental and human systems (Latifovica et al. 2005; Geist et al. 2006). Factors such as low soil pH, harvesting of afforested areas and other restoration related construction activities (Bröring and Wiegleb 2005) acted together or independently to bring about the LCCs.

AUTHORS' CONTRIBUTIONS

EKA – Was the brain behind the study. Participated in writing the introduction, methods, results and discussion parts of the manuscript. Participated the in revision of the manuscript after receiving reviewers' comments. JBD and GW – Participated in writing the introduction, methods, results and discussion part of the manuscript. Participated the in revision of the manuscript after receiving reviewers' comments. SBA – Supported the analyses of land use changes and presenting of results. He contributed in providing relevant inferences of land cover changes in the post-mining landscape. Participated the in revision of the manuscript after receiving reviewers' comments. KT – Provided technical advice and gave significant contribution during the writing of the manuscript. Participated the in revision of the manuscript after receiving reviewers' comments. All authors read and approved the final manuscript.

REFERENCES

1. Antwi EK, Boakye-Danquah J, Asabere SB, Yiran GAB, Loh SK, Gyekye AK, Abagale FK, Asubonteng KO, Attua EM, Barimah AO (2014) Land Use and Landscape Structural Changes in the Ecoregions of Ghana. Journal of Disaster Research 18(4):452-467

2. Antwi EK, Krawczynski R, Wiegleb G (2008) Detecting the effect of disturbance on habitat diversity and land cover change in a post-mining area using GIS. Landscape Urban Plan 87:22-32 doi: 10.1016/j.landurbplan.2008.03.009
3. Blumrich H, Bröring U, Felinks B, Fromm H, Mrzljak J, Schulz F, Vorwald J, Wiegleb G (1998) Naturschutz in der Bergbaufolgelandschaft -Leitbildentwicklung. Studien und Tagungsberichte 17:1-44
4. Bradshaw AD (1999) The Importance of Nitrogen in the Remediation of Degraded Land. In: Wong MH, Raton JWC (eds) Reclamation and Management of Degraded Lands. Fla, pp. 153-162
5. Bröring U, Wiegleb G (2005) Soil zoology II: colonization, distribution, and abundance of terrestrial heteroptera in open landscapes of former brown coal mining areas. Ecol Eng 24:135-147 doi: 10.1016/j.ecoleng.2004.12.015
6. Bungart R, Hüttl RF (2001) Production of biomass for energy in post-mining landscapes and nutrient dynamics. Biomass Bioenerg 20(3):181-187 doi:10.1016/S0961-9534(00)00078-7
7. Cohen WB, Goward SN (2004) Landsat's role in ecological applications of remote sensing. Bioscience 54(6):535-545 doi:10.1641/0006-3568(2004)054[0535:LRIEAO]2.0.CO;2
8. Demirel N, Düzgün , Emil MK (2011) Landuse change detection in a surface coal mine area using multi-temporal high-resolution satellite images. Int J Mining Recl Env. 342-349 doi:10.1080/17480930.2011.608889
9. Der Braunkohlenausschuss (1994) Sanierungsplan Schlabendorf Felder. Gesetz - und Verordnungsblatt für das Land Brandenburg. Teil II – Nr. 14 von 8.
10. Drummond MA, Loveland TR (2010) Land-use pressure and a transition to forest-cover loss in the Eastern United States. Bioscience 4(60):11-12 doi:10.1525/bio.2010.60.4.7
11. Edenius L, Vencatasawmy CP, Sandstrom P, Dahlberg U (2003) Combining Satellite Imagery and Ancillary Data to Map Snowbed Vegetation Important to Reindeer Rangifer Tarandus. Arct Antarct Alp Res 35(2):150-157 doi:10.1657/1523-0430(2003)035[0150:CSIAAD]2.0.CO;2

12. ERDAS (2002) Erdas Field Guide, Version 8.6.
13. Felinks B (2000) Primärsukzession von Phytozönosen in der Niederlausitzer Bergbaufolgelandschaft. Cottbus.
14. Fernández V, Eichert T (2009) Uptake of hydrophilic solutes through plant leaves: Current state of knowledge and perspectives of foliar fertilization. Crit Rev Plant Sci 28:36-68
15. Geist H, McConnell W, Lambin EF, Moran E, Alve A, Rudel T (2006) Causes and Trajectories of Land-Use/Cover Change. In: Lambin EF, Geist HJ (eds) Land-Use and Land-Cover Change: Local Processes and Global Impacts. The IGBP Series, Berlin Heidelberg: Springer-Verlag.
16. Gould W (2000) Remote sensing of vegetation, plant species richness, and regional biodiversity hotspots. Ecol Appl 10:1861-1870 doi:10.1890/1051-0761(2000)010[1861:RSOVPS]2.0.CO;2
17. Groninger J, Skousen J, Angel P, Barton C, Burger J, Zipper C (2007) Mine Reclamation Practices to enhance Forest Development through Natural Succession. Forest Reclamation Advisory No. 5. http://www.prp.cses.vt.edu/ARRI/FRA-No5.pdf *webcite*. Accessed 20 Oct 2010
18. Grünewald U (2001) Water resource management in river catchments influenced by lignite mining. Ecol Eng 17:143-152 doi:10.1016/S0925-8574(00)00154-3
19. Haigh MJ (ed) (2000) Reclaimed Land: Erosion Control, Soil and Ecology A. A. Balkema, Rotterdam.
20. Hendrychová M (2008) Reclamation success in post-mining landscapes in the Czech Republic: a review of pedological and biological studies. J Landscape Stud 1:63-78
21. Hill MO, Gauch HG (1980) Detrended correspondence analysis: an improved ordination technique. Vegetatio 42:47-58
22. Hüttl RF (1998) Ecology of Post-Mining Landscapes in the Lusatian Lignite Mining District, Germany. In: Fox HR, Moore HM, McIntosh AD (eds) Land Reclamation – Achieving Sustainable Benefits, Rotterdam: A. A. Balkema. pp 187-192
23. Knoche D, Embacher A, Katzur J (2000) Element dynamics of oak ecosystems on acid-sulphurous mine soils in the lusatian lignite mining district (Eastern Germany). Landscape Urban Plan 51:113-122 doi:10.1016/S0169-2046(00)00102-X

24. Latifovica R, Fytasb K, Chenc J, Paraszczakb J (2005) Assessing land cover change resulting from large surface mining development. Int J Appl Earth Obs Geoinf 7:29-48 doi:10.1016/j.jag.2004.11.003
25. Nova Scotia Environment (2009) Guide For Surface Coal Mine Reclamation Plans. Policy and Corporate Services Division Environmental Assessment Branch. pp 1-17
26. Palmer M (2010) Detrended Correspondence Analysis. http://ordination.okstate.edu/DCA.htm *webcite*. Accessed 20 Aug 2010
27. Rahdary V, Soffianian A, Najfabdai S, Khajeddin SJ, Pahlavanrav I (2008) Land use and land cover change detection of mouteh wildlife refuge using remotely sensed data and geographic information system. World Applied Sci J 3(1):113-118
28. Raumann CG, Cablk ME (2008) Change in the forested and developed landscape of the Lake Tahoe Basin, California and Nevada, USA, 1940–2002. Forest Ecol Manag 255:3424-3439 doi:10.1016/j.foreco.2008.02.028
29. Ravan SA, Roy PS, Sharma CM (1998) Accuracy evaluation of digital classification of Landsat TM data. An approach to include phenological stages of tropical dry deciduous forest. Int J Ecol Env Sci 22:33-43
30. Schultz F, Wiegleb G (2000) Development options of natural habitats in a post-mining landscape. Land Degrad Dev 11:99-110 doi:10.1002/(SICI)1099-145X(200003/04)11:2 <99::AID-LDR368>3.0.CO;2-I
31. Swangjang K, Iamaram V (2011) Change of land use patterns in the areas close to the airport development area and some implicating factors. Sustainability 3:1517-1530 doi:10.3390/su3091517
32. Ter Braak CJF (1996) Unimodal Methods to Relate Species to Environment. Centre for Biometry Wageningen (DLO Agricultural Mathematics Group). Wageningen, the Netherlands. p 266
33. Ter Braak CJF, Smilauer P (2002) CANOCO Reference Manual and CanoDraw for Windows User's Guide: Software for Canonical Community Ordination (version 4.5). Microcomputer Power, Ithaca, New York, USA.

34. Weber D, Hintermann U, Zangger A (2004) Scale and trends in species richness: considerations for monitoring biological diversity for political purposes. Glob Ecol Biogeogr 13:97-104 doi:10.1111/j.1466-882X.2004.00078.x
35. Wiegleb G, Felinks B (2001) Primary succession in post-mining landscapes-chance or necessity? Ecol Eng 17:199-217 doi:10.1016/S0925-8574(00)00159-2
36. Wiegleb G, Felinks B (2001) Predictability of early stages of primary succession in post-mining landscapes of lower lusatia. Appl Veg Sci 4:5-18 doi:10.1016/S0925-8574(00)00159-2
37. Wollum AG (1994) Soil sampling for microbiological analysis. In: Methods of soil analysis, Part 2, microbiological and biochemical properties, SSSA Book Series No. 5. Soil Science Society of America, Madison, WI. pp 2-13
38. Yang JJ (2008) Detecting landscape changes pre-and post surface coal mining in Indiana, USA. Ann GIS 14:1 doi:10.1080/10824000809480637
39. Young SS, Harris R (2005) Changing patterns of global-scale vegetation photosynthesis 1982–1999. Int J Remote Sens. doi:10.1080/01431160500239198 http://eprints.ucl.ac.uk/1065/1/1065.pdf

Chapter 3

Towards Safety from Toxic Gases in Underground Mines Using Wireless Sensor Networks and Ambient Intelligence

Isaac O. Osunmakinde

Semantic Computing Group, School of Computing, College of Science, Engineering and Technology, University of South Africa, (UNISA), Pretoria 0003, South Africa

ABSTRACT

The growing number of fatalities among miners caused by toxic gases puts pressure on the mining industry; innovative approaches are required to improve underground miners' health. Toxic gases are very often released in underground mines and cannot easily be detected by human senses. This paper investigates the presence of the inherent types of toxic fumes in critical regions and their suspension and trends in the air and intends to generate knowledge that will assist in preventing miners from contracting diseases. The development of intelligent decision support systems is still in its infancy. Knowledge of how to make them profitable in improving miners' safety is largely lacking. An autonomous remote monitoring framework of wireless sensor networks, which integrates mobile sensing and Ohm's law, coupled with ambient intelligence governing decision-making for miners, is developed. The framework has been investigated in indoor scenarios and successfully deployed for real-life application in an aeronautic engine test cell environment, such as those typically found in underground mines. Useful demonstrations of the system were carried out to provide similar knowledge to safeguard engineers from the inhalation of toxic gases. This provides early warning for safety agents. The system has proven to be suitable for deployment in underground mines.

INTRODUCTION

Efficient monitoring of concentrations of toxic gases in restricted areas, such as those typically found in mines, is regarded as a problem of special significance by the mining industry [1]. This challenge is significantly greater in an unstructured underground mining environment, such as a stope. A mine stope is formed as a result of repeatedly blasting hard rock from strategic sides with explosives. After mine blasting or other mining activities, such as drilling, construction, and barring, the underground terrains become restricted, unknown, unstructured, and particularly dangerous owing to the presence of poisonous explosive gases. The mine gases are highly concentrated and toxic, become a threat to both miners' health and the environment, and also limit visibility.

Figure 1 shows an example of a buildup and explosion of toxic gases, such as methane (CH_4), nitrogen dioxide (NO_2), and hydrogen sulphide (H_2S) in an underground mine. The inhalation of an overlooked black damp in mines, which is a mixture of dust, carbon dioxide, and nitrogen [2], gradually overcomes and destroys the body's blood system, which may result in chronic diseases as a consequence of exposure to toxic gases. Research conducted in 2006 revealed a significantly elevated risk of leukaemia among German employees with a long occupational career in underground mines [1]. Several recent references to the incidence of mining-related diseases can be found in [3, 4].

Figure 1: A sample underground mine explosion due to the buildup of toxic gases.

It is therefore worth noting that, after blasting, miners currently have to wait outside the mines for a specified period of time for the poisonous gases/dust to dissipate. The question remains whether the gas/dust disappears completely in the air or merely reduces below that which the human nose can sense before miners enter the unsafe areas. Understanding of the density of the poisonous gases suspended in the air will assist in preventing miners from contracting dreaded diseases and is manifestly a sound basis for improving safety procedures in mines.

Predominant current efforts to monitor gas levels are focused on the use of [5] (i) lag/waiting time, (ii) mining ventilation, (iii) preparedness, and (iv) government regulations and agencies such as the Clean Air Act and Environmental Protection Agency in the USA. These promote activities that reduce air pollution. Each of these practices is successful for some environmental monitoring applications, but they require model intelligence. However, other efforts are being made to combat the problems of toxic fumes, such as the traditional fixed pollution monitoring stations measuring the air quality index, for example, the sophisticated equipment setups in Swansea [6] and in Ohio, USA [7]. They are highly sensitive and well calibrated, but the systems are operated with poor flexibility and expansibility, and the stations are expensive, which could be unaffordable for many developing mines and countries.

We seek a research solution to the deficiencies in the current monitoring approaches, which necessitate real-time sensing and decision-making through a wireless sensor network (WSN). WSNs are characterised by self-organisation, wireless communication, distributed, autonomous, and simple maintenance, which have applications in the research of environmental and underground monitoring systems [8]. This research focuses on making useful decisions on improving safety in restricted areas. The areas contain a roughly dynamic compartment whose sources of toxic gases have to be identified and remotely monitored for safety. Relatively little research has focused on providing methods for knowledge generation in WSNs that can support real-time collaborative decision-making on improving underground safety. The prevailing approaches often use structured query language- (SQL-) like primitives, and events are defined using a subevent list and confidence functions in SQL [9]. However, SQL is not appropriate for describing gas-sensing events in WSNs, as it (i) cannot capture data dependencies and interactions among different sensing scenarios, (ii) does not really support probability models, (iii) is inappropriate in describing complex temporal constraints and data dependencies, (iv) lacks the ability to support collaborative decision-making and triggers, and (v) does not facilitate any global analysis of the gas-sensing event system. In our previous research [10], a Bayesian belief network was used for environmental situation recognition in WSNs. The network was used in offline mode as a main representational structure of a ubiquitous sensor network, which only uses qualitative analysis for intelligence. However, we believe that crisp values for qualitative knowledge cannot adequately handle the often imprecise sensor readings. In this paper we demonstrate that using both qualitative and quantitative reasoning in real time significantly improves the knowledge generated.

In mining environments, network routing efficiency in WSNs stands out as a pivotal factor [11] beside other research problems. Recently, modelling the complexities of underground tunnel communication systems has been a key focus and has been extensively studied in [12, 13]. Long-distance WSN in [14] was applied in a coal mine by building on an integrated mine network. It was proved that it is suitable for a mine environment. We have studied different existing real-world scenarios where WSNs are being applied. Based on this study we have discovered that there are significant commonalities, but we do not know of any methodology that (i) provides detail of implementation for industry reproducibility and (ii) specifies the best practices that should be used in general, from the level of gas sensing up to decision-making in the case of this restricted area. As recommended in [15], newer frameworks need the extension of technologies to monitor complex toxic situations effectively. A real-time remote monitoring framework of WSNs and ambient intelligence based on temporal statistical methods governing decision-making for miners is developed for bridging the challenging gaps. The major contributions of this paper are as follows:

- the development of a reactive remote monitoring framework based on real-time gas sensing, which spans physical layers through application space, by building on mine networks;
- the modelling of the theory in WSNs, Ohm's law, concepts of static and mobile robot sensing, derivations of curvilinear gas calibration equations, and ambient intelligence based on temporal statistical methods;
- the applications to monitor gases remotely in a real-life engine test cell, such as those typically found in underground mines, generating knowledge to improve safety with an assurance of assistance whenever required.

The rest of the paper is arranged as follows: Section 2 presents the background, which includes toxic mine gases, their physiological health effects, and the baseline mine network; Section 3 presents the proposed remote monitoring framework, which includes the sensor publishers, mathematical equations for gas calibration, and the ambient intelligence approach; Section 4 presents two experimental field setups; Section 5 critically presents experimental and comparative evaluations together with benchmarking of our proposed framework. We conclude the paper in Section 6.

BACKGROUND

This section presents toxic mine gases, their physiological health effects, and the baseline mine network. The purpose is to provide the necessary knowledge to understand the requirements of the newly proposed framework fully.

Toxic Mine Gases and Physiological Health Effects

Table 1 presents some information about toxic gases commonly found in mine air. The knowledge is extracted from [16]. This information benchmarks our intelligence system to guide the early warning triggers.

Table 1: Common mine gases

Gases	Properties	Health effects	Concentration in the air
CO2	Colourless; odourless; heavier than air; acidic taste at high concentrations	At 5%, stimulated respiration; at 7% to 10%, unconsciousness after few minutes of exposure	280–390 ppm
CO	Flammable; colourless; tasteless; odourless; lighter than air	At 200 ppm, slight headache, tiredness, dizziness, nausea after 2 to 3 hrs; at >200 ppm, life threatening after 3 hrs	0-trace ppm

H2	Colourless; reacts easily with other chemical substances; explosive mixtures are easily formed; lighter than air	High concentration causes oxygen-deficient environment; headaches; ringing in ears; drowsiness; nausea; skin having blue colour	0.5 ppm
CH4	Colourless; odourless; tasteless; flammable; lighter than air; largest component of fire damp	Asphyxiation, dizziness, headache, and nausea in high concentrations due to displacement of oxygen	1.79–2.0 ppm
NO2	Reddish-brown colour in high concentrations; acrid or bleach odour; nonflammable; heavier than air	At 1–13 ppm, irritation of nose and throat; ≤80 ppm,tightness in chest after 3 to 5 minutes; >80 ppm, pulmonary edema after 30 minutes	0.03 ppm

Baseline Mine Network for Remote Monitoring

The integrated mine network proposed in [23] is intended for use as a baseline to connect with the proposed framework. The integrated mine network is composed of optical fibre as the main network through the shaft and uses WSNs in the roadwaywheremonitoring is needed. TheWSNs also connect other networks via gateway nodes, realising information interaction of aboveground and underground networks, as shown in Figure 2. The benefits of this mine network are as follows: (1) full usage of the existing network in the mine, which implies low maintenance, (2) no need for a WSN to lay lines for communication and power, whichmakes itmore flexible, (3) sensor nodes that could be positioned optionally and compactly, making it possible to monitor stopes and any other areas in the mine, and (4) expandability of the network even when nodes and roadways are modified, making it convenient to monitormoving robot sensing in underground mines.

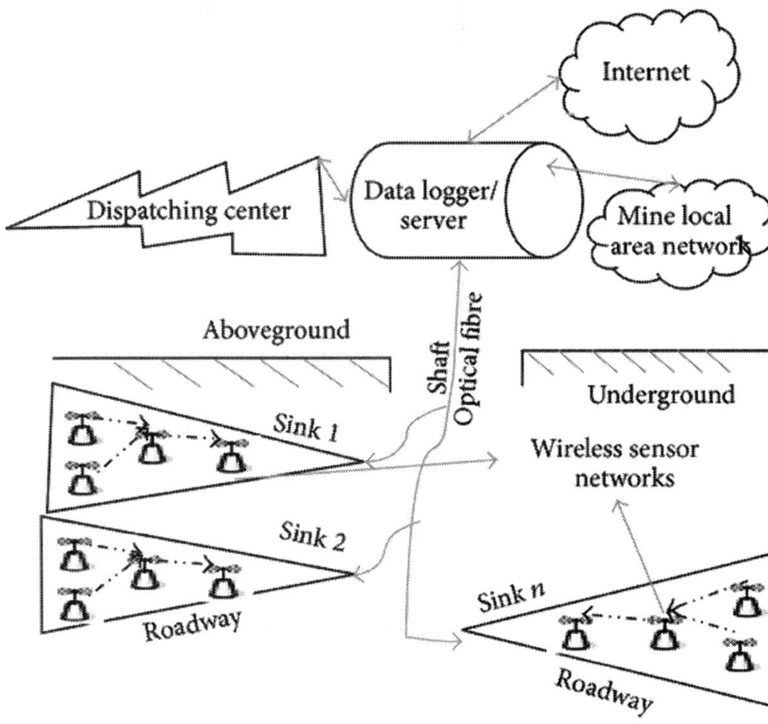

Figure 2: Integrated mine network.

In our approach, the network is intended for deploying the sensing nodes to identify toxic gases and sense what is happening underground. The inclusion of fibre-optic cabling through the shaft alleviates the challenges of no network coverage in underground tunnel regions. The fibre-optic cable communicates to a server or data-logger aboveground, which stores sensor readings for forward transmission to the Internet or network space, as used in Figure 3.

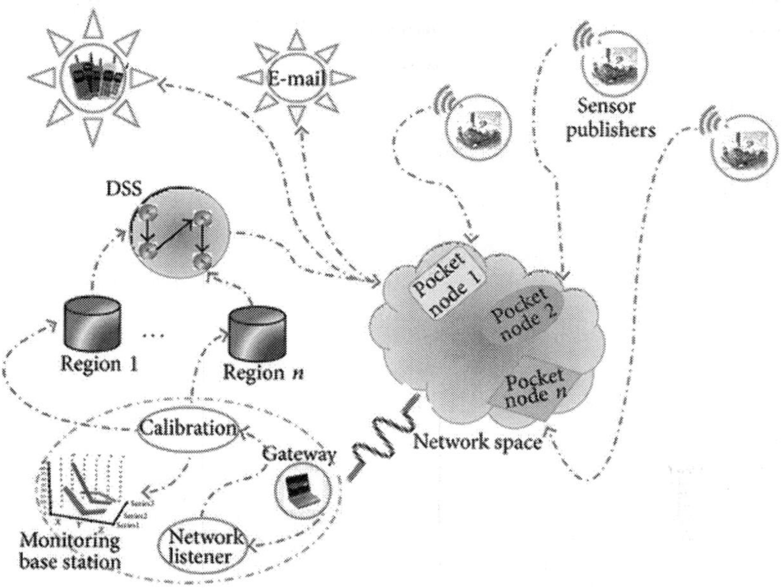

Figure 3: New remote gas monitoring framework.

PROPOSED REMOTE MONITORING GAS SENSING FRAMEWORK

This section introduces the new reactive framework in Figure 3 for remotely monitoring toxic gases in restricted areas, such as those typically found in underground mines. It uses the WSN and ambient intelligence technology by building on Figure 2. The goal of this framework is to guide the monitoring of toxic gases in restricted areas through a set of well-defined nodes in its entire real-time system. As shown, the reactive system is divided into three major nodes: sensor publishers in Section 3.1, gas calibration in Section 3.2, and intelligent decision support system (DSS) nodes in Section 3.3. We expect the benefits of this framework to be numerous: (i) mobile robot sensing balancing between maximising the detection rate of gases and minimising the number of missed detections/unit areas, (ii) the timely detection and early warning of many common problems such as failing sensors, toxic fume levels hazardous to health in various regions, and potential gas explosions, (iii) an intelligent goal-directed decision-making process, (iv) an increase in coordination between the different working nodes on the framework, thanks to a clear division of responsibilities, and (v) the reproducibility of knowledge gathered from real-world WSN applications.

Sensor Publisher Nodes-Physical Layer

The sensor publishers, implemented as shown at the top right of Figure 3, collect gas data from different regions of an environment, which is necessary to understand the concentrations of the gases being monitored fully. This physical layer consists of an exhaustive design of the framework, which could include mobile sensing, shown in Figure 4(a). This is expatiated in the following.

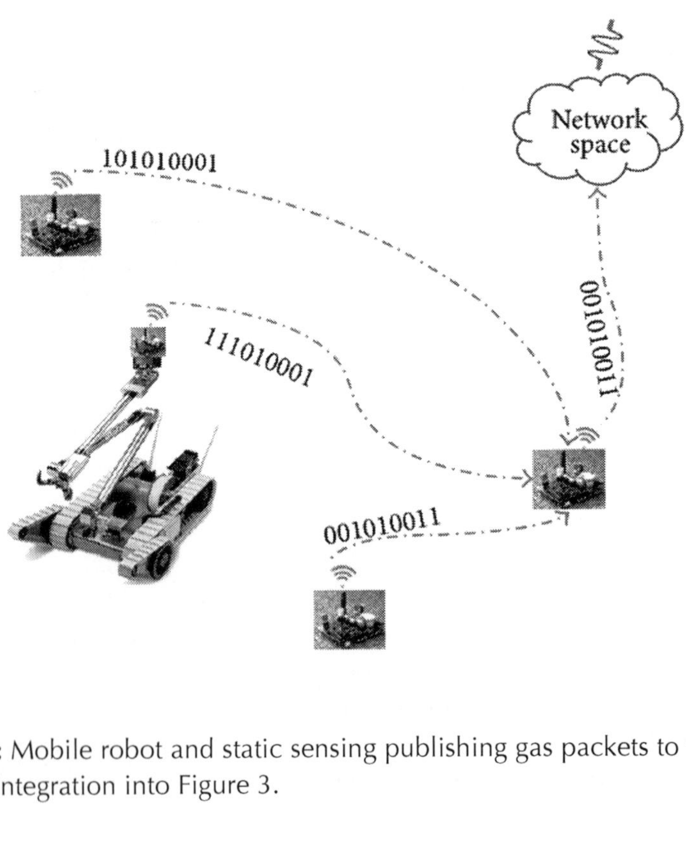

Figure 4: Mobile robot and static sensing publishing gas packets to the gateway for integration into Figure 3.

Static and mobile robot sensing nodes provide an electronic "nose" on board, which includes an SD card slot and GPRS capabilities for sending an sms, and come in different communication flavours of Zigbee using the Xbee protocol. Zigbee can operate on line-of-sight or use the direct-sequence spread spectrum to penetrate barriers. The node uses a lithium battery, which can be recharged through a dedicated socket by a solar panel, allows extra sensors and GPS on board, and can transmit over a distance of 75?m. Low-level software programming is embedded onto different board circuitry, enabling wireless communication. Tagging each sensor board at this physical layer enables localisation of the source of various toxic gases, thereby making concurrent gas publishing onto the network space identifiable in Figure 4(b) as separate packet voltage (v) values. Thus, the network space aligns our framework with the integrated mine network, where the network space represents the Internet.

Mathematical Equations for Gas Calibration Nodes-Middle Layer

Every sensor used in gas measurement can be characterised by a specific response function or calibrationequation relating the sensor output voltage ($\boxtimes$) to a gas concentration normallymeasured in parts permillion (ppm). The goal of thismiddle layer is to input the voltage and empirically generate the ppm that subsequent nodes in Figure 3 use. The raw gas measurements captured over the sensors are usually in millivolts (mV), and they are converted to $\boxtimes$, obviously by dividing by 1,000. This needs to be converted back to the resistance of the sensor using Ohm's law [24] in

$$R_s = \frac{V}{I} = \frac{V}{V_s/R_l} \Longrightarrow \frac{V * R_l}{V_s}, \qquad (1)$$

where $\boxtimes_{\boxtimes}$ is the resistance of the sensor, $\boxtimes$ is the voltage supplied, $\boxtimes$ is the current over the sensor, $\boxtimes_{\boxtimes}$ is the constant load resistance, and $\boxtimes_{\boxtimes}$ is the voltage over the sensor.When a sensor remains without power for a prolonged period of time, it possibly shows an unstable output. This stability is regained after many consecutive cycles of power supply with a defined $\boxtimes_{\boxtimes}$ over the sensor, leading the resistance of the sensor to

$$\Longrightarrow R_s = \frac{V * R_l}{(V_s - R_l)}. \tag{2}$$

More information and the values of $⊠_⊠$ and ⊠ can be found in gas datasheets in [25].Therefore calibration of the gas sensors involves two steps,which are (i) deriving base resistance (zero point) and (ii) span calibration. Since there is no established standard that defines zero air, a good reference point can be an office area where air may be considered fresh. The sensors used here are relative; no device gives the same reading for a certain concentration of gas, but the relationship between the readings for a particular sensor for different concentrations of readings is constant. About ten readings could be experimentally taken in fresh air and averaged to give the calibration point ($⊠_0$) of the gas sensor. For the span calibration, subsequent readings using (2) will be compared to ⊠0 to form a ratio $⊠_⊠ /⊠_0$. The calibration node collects mathematical equations (3)–(9) and autonomously finds a well-fitted model for translating the ratio to corresponding ppm, depending on the known nature of the behaviour of the gas being measured and the measurement process itself.This process minimises any sensor inaccuracies by regulating the reliability of its readings.

The linear regression model is expressed in (3) as

$$\widehat{y} = m * r + c, \tag{3}$$

Where

$$m = \frac{(n \sum xy - \sum x \sum y)}{\left(n \sum x^2 - (\sum x)^2\right)},$$

$$c = \overline{y} - m * \overline{x}, \tag{4}$$

$\widehat{Y}$ is the estimated ppm, ⊠ is the ratio, and ⊠ and ⊠ are known volts and ppm. The goodness of fit of the model is examined using the Pearson correlation coefficient, ⊠, in

$$R = 1 - \frac{\sum_i ((x_i - \overline{x})(y_i - \overline{y}))}{\sqrt{\left(\sum_i (x_i - \overline{x})^2\right)\left(\sum_i (y_i - \overline{y})^2\right)}}. \tag{5}$$

If ⊠ and ⊠ are linearly correlated, ⊠ would be strongly close to +1 or −1, then the system selects (4) for the calibration. Otherwise, any of the four curvilinear models in (6)–(9) is selected, depending on the highest value of ⊠:

Power series:

$$y = ax^b, \tag{6}$$

Logarithmic:

$$y = \ln\left(ax^b\right), \tag{7}$$

Exponential:

$$y = ae^{\alpha x}, \tag{8}$$

Polynomial:

$$y = ax^2 - bx + \beta. \tag{9}$$

Parameters ⊠, ⊠, ⊠, and ⊠ are estimated by linearising equations (6)–(9), which can be found in [26]. In the present case, the system selected is (i) power series for calibrating gases CO, NO_2, H_2, CH_4, and NH_3, (ii) polynomial model for calibrating H_2S, and (iii) exponential for calibrating CO_2.

Ambient Intelligence Node for Decision Support System-Application Layer

Sensors are generally believed to be imprecise and not completely accurate even after the calibration process. To increase our confidence in the presence of uncertainty in toxic gas sensing, an intelligent DSS for reasoning over some periods of time is needed, as shown in Figure 3. One of the important tasks in our remote gas monitoring is sending early warning to safety officers through the real-time intelligence node based on temporal statistical models for reasoning on the imprecise gas sensor readings and allowing users interaction through the theory of situation awareness (SA) [10, 27]. We prefer the theory and the models because of ease in the interpretation of results, fast execution time, and scalability for large multidimensional data sets.

The objective of the SA was to guide safety officers through the decision-making process by detecting hidden patterns in the data captured over the WSN and revealing what is happening, why it is happening to the environment, what can be done to avoid unwanted behaviour, and what will happen next. In consolidating SA, we construct the temporal statistical mean, median, mode, standard deviation, and skewness in Table 2 for describing the situation of the toxic gases, while chi-square inference based on contingency table analyses relationships between parameters in multiple regions over time.

Table 2: Temporal statistical methods used in real time.

Temporal statistical methods	Their models
Mean ($\overline{\chi_t}$) and standard deviation (SD)	$\overline{\chi}_t = \frac{1}{n}\sum_{i=1}^{n}\chi_1;$;
	$SD_t = \sqrt{\frac{1}{n-1}\Sigma^n{}_{i=1}\left(\chi - \overline{\chi}\right)^2}$
Skewness (St)	$SD_t = \frac{\Sigma^n{}_{i=1}\left(\chi_i - \overline{\chi}\right)^3}{(n-1)s^3}$

As the toxic gases are observed over time, the statistical models change dynamically in values describing the location and variability of situations based on what is intended to be known in the regions observed. Since skewness measures the shapes of distributions, it characterises the degree of asymmetry of a distribution around its mean [28]. This means that, when skewness > 0 or < 0, the gas concentrations are clustered on one side, and when skewness = 0, they are distributed normally.

For analysing the relationships between gases, the following five stages are required: (i) a contingency table, which is a frequency between two ormore variables, is first constructed, while independence between the variables is conducted by computing the expected values from the observed values using (10); (ii) a chi-square test and degree of freedom are then computed as shown in (11) and (12), respectively; (iii) a temporal probability (⊠⊠) representing the degree of independence is computed using an ⊠ 2distribution function as shown in (13); (iv) an inference on the relationship is made by comparing the ⊠⊠ with a significant level often set at 5%; and (v) if ⊠⊠ < 0.05, the two variables thus have a relationship, otherwise no relationship can be concluded in the contingency table as follows:

$$e_{ij} = \frac{\sum_i o_{ij} * \sum_j o_{ij}}{\sum_i \sum_j o_{ij}}, \quad (10)$$

$$\chi_t^{\ 2} = \sum_i \sum_j \frac{\left(o_{ij} - e_{ij}\right)^2}{e_{ij}}, \quad (11)$$

$$df_t = (\text{total rows} - 1) * (\text{total columns} - 1), \quad (12)$$

$$p_t = \chi^2\text{DIST}\left(\chi^2, df\right) \quad (13)$$

It is believed that the theory of SA and the temporal statistics can significantly minimise the number of false positives, as they combine qualitative and quantitative reasoning. To the best of our knowledge, no previous work on applying decision-making to gas monitoring has considered the effects of both theories on the accuracy of early warning. The contribution of this phase is an integration of the set of calibrated gases from multiple zones and building the temporal statistical models for real-time decision-making. The output of the ambient system is the provision of accurate early warning for each zone, taking into account cell phone and E-mail communication with the safety officers.

EXPERIMENTAL FIELD SETUPS

Hypothesis 1. Mobile sensing will not track the variability of gas concentration in air as well as static sensing over WSNs indoors.

In other words, it might not be necessary to implement mobile and static gas sensing over WSN indoors if it is known ahead of time that an environmental volume of air cannot be polluted beyond absorbable levels. To test the hypothesis, it is necessary first to establish different fields' setups, such as Figure 5(a) showing a cross-ventilation area and Figure 5(b) with no free flow of fresh air.

(a)

(b)

Figure 5: (a) Static sensing node 1 on a filing cabinet in an aerated office, 5 metres away from the monitoring station. (b) Sensing node 2 on a mobile pioneer robot at the printing and network server section, 10 metres and two doors (barriers) away from the monitoring station.

Hypothesis 2. The performance of remote monitoring of gases in restricted areas over WSNs is not correlated to its performance in areas such as those typically found in underground mines.

An underground mine can be modelled in a more or less detailed fashion as a restricted area. For a mine consisting of multiple features such as wall barriers, the features would ideally need to be modelled separately. A real-life setup in two aeronautic mechanical engine test cells with dimensions 800?m^3 and 650?m^3 was used to test this hypothesis. Although it will not be able to reveal whether more accurate setups give better results in all cases, it will allow us to observe the effects of wall barriers encountered during transmission of gas packets outside the enclosed test cell, such as those typically found in underground mines. The good news is that our demonstrator does not rely only on line-of-sight for transmission, but also uses a direct-sequence spread spectrum to penetrate the barriers.

To assess the readiness of our system for underground mine deployment in terms of sensing some toxic gases and localising their sources, the following objectives are set: (i) performing ambient measurements assessing temperature, CO_2, NO_2, and CH_4 buildup or distributions, (ii) getting a feel for the gas flow, (iii) picking up other possible fumes, and (iv) responding to environmental activities/events in the cell. Because of confidentiality in industries, the real-life engine test cell is presented as shown in Figure 6. A sensor node $\boxtimes_1$ is located high up, close to the inlet, where fresh air enters the cell, a sensor node $\boxtimes_2$ is located at the basement in front of the engine, and a sensor node $\boxtimes_3$ is located behind the engine close to the air extractor, while the base monitoring station is outside the cell.

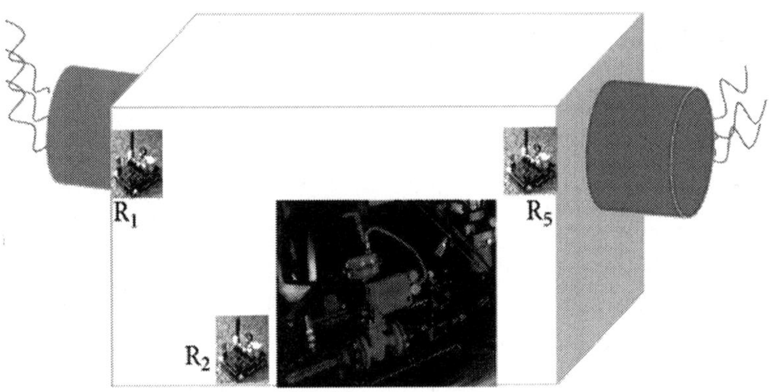

Figure 6: A mechanical engine test cell, over 20 metres and four walls (barriers) away from Figure 4(b).

EXPERIMENTAL RESULTS AND EVALUATIONS

From the ambient measurements, the six variables captured are quantitative data types where temperature is the warmness of the region measured in degrees Celsius (°C), and region values are in integers, while gases CO_2, CH_4, and NO_2 are in ppm. The time step on the $\boxtimes$-axis of Figures 8 and 9 is in seconds, and the interval between two time step units was set to 40 seconds. The last time step 80 on the figures implies that the sampling time or observations in the test cell took about one hour. The data model used in these experiments is in the form shown in Figure 7.

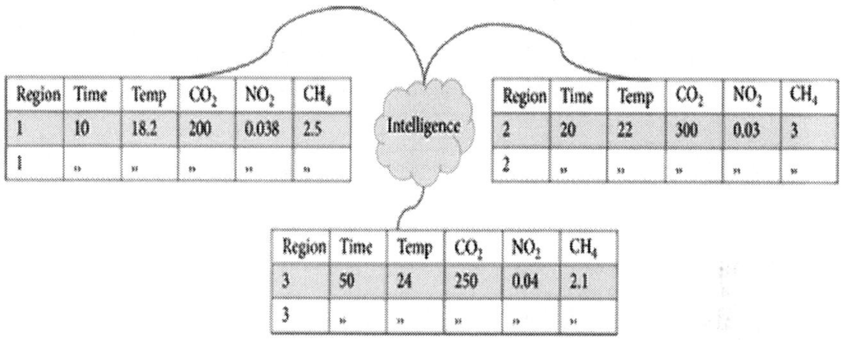

Figure 7: Sample gases data model and the ambient intelligence.

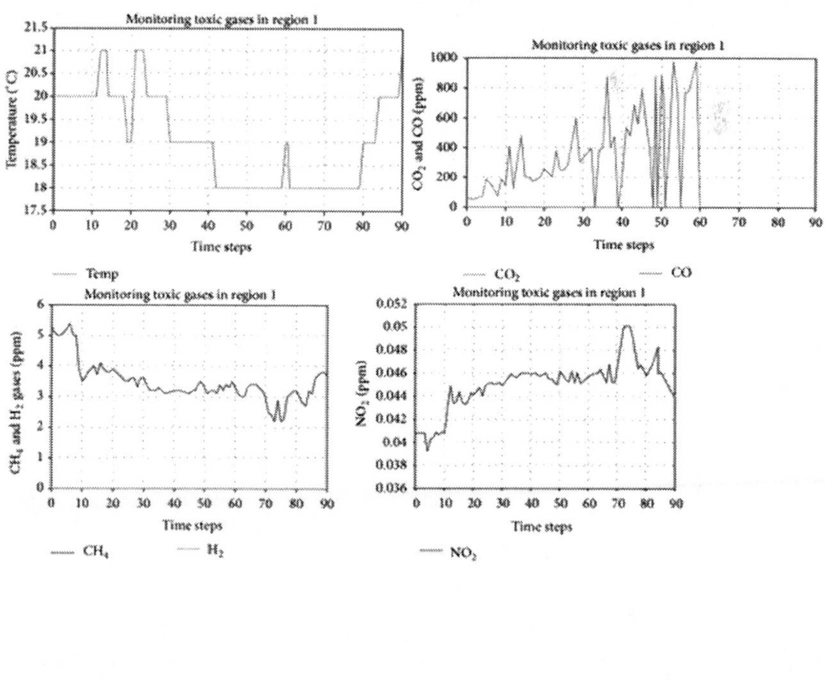

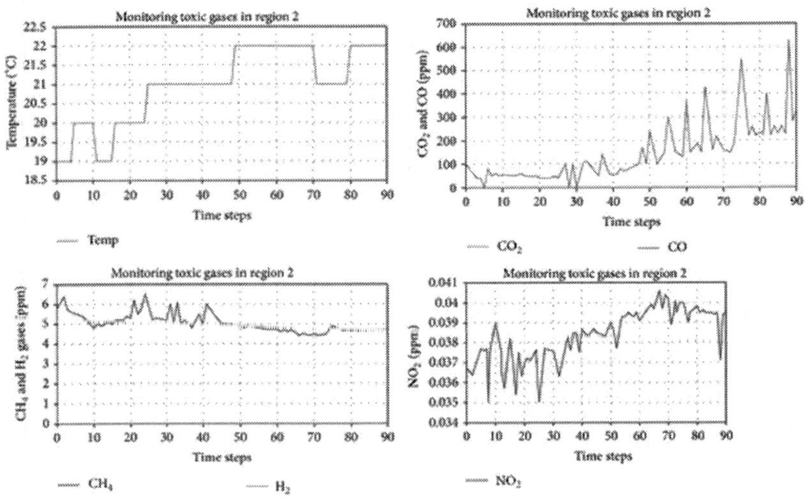

Figure 8: (i) Temperature at the top left-hand corners, (ii) CO_2 and CO expected at the top right-hand corners, (iii) CH_4 and H_2 expected at the bottom left-hand corners, and (iv) NO_2 at the bottom right-hand corners of (a) and (b).

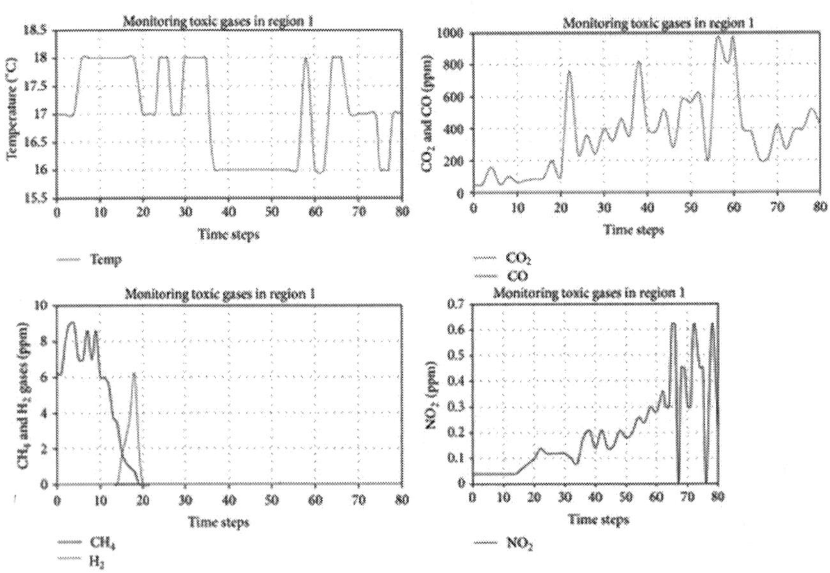

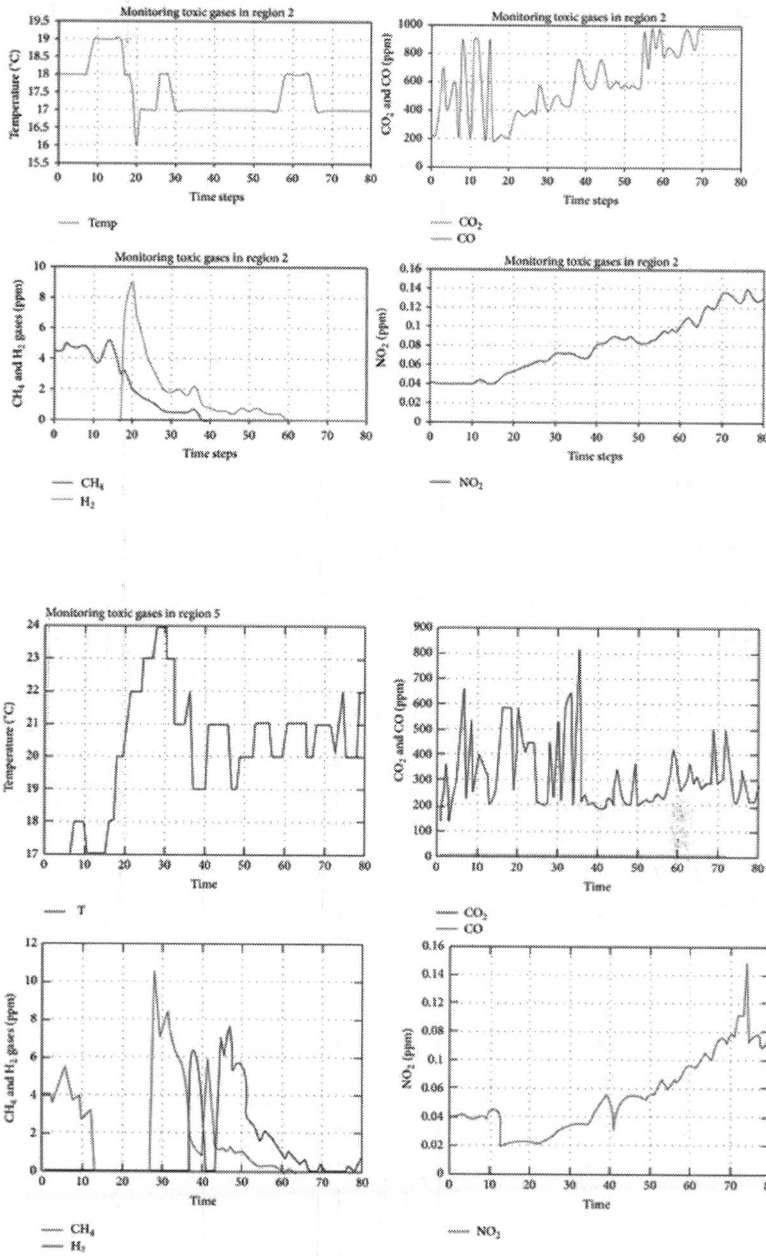

Figure 9: (i) Temperature at the top left-hand corners, (ii) CO_2 and CO expected at the top right-hand corners, (iii) CH_4 and H_2 expected at the bottom

left-hand corners, and (iv) NO_2 at the bottom right-hand corners of (a), (b), and (c).

Hypothesis 1 (indoor environments). The static and mobile sensor nodes in Figure 5 capture the concentration of gases in the air, publish a stream of gases in voltages over the WSN onto a gateway subscription, and calibrate it into ppm, as shown in Figure 4(b).

Situation Awareness for Indoors-Qualitative Reasoning

The objective here is to assess the qualitative performance of the static sensing in a ventilated environment with the region 1 results shown over time in Figure 8(a), which exhibits noticeable differences when compared to the mobile sensing in an enclosed environment with region 2 results in Figure 8(b). In particular, the SA questioning paradigm in the following guides the safety officers through a decision-making process.

Decision-Making: Sample Qualitative Reasoning at Time Step 90.?:

? Q_1: *What is happening?*

? A_1: CH_4 concentration and temperature seemslightly higher in $\boxtimes_2$ than $\boxtimes_1$; higher instability of CO_2 in $\boxtimes_1$ than $\boxtimes_2$; slightly more NO_2 concentration in $\boxtimes_1$ than $\boxtimes_2$.

? Q_2: *Why is it happening?* A_2: Higher fresh air and noise dilution in $\boxtimes_1$ than $\boxtimes_2$; suspicious clearer spikes in $\boxtimes_2$ due to steady mobility and of course machines in $\boxtimes_2$ generate more heat.

? Q_3: *What can we do about it?* A_3: Unwanted behaviours could be avoided if detonation of explosives is avoided, and machines are serviced regularly.

? Q_4: *What will happen next?*

? A_4: Since the concentrations approximately revolve around the acceptable values shown in Table 1, the two regions seem relatively safe to health.

Hypothesis 1 states that mobile sensing will not track the variability of gas concentration in air as well as static sensing over WSNs indoors. The qualitative results, however, show that it is not the case, implying that mobile sensing accurately captures the dynamism in region 2, as well as static sensing does in region 1.

Hypothesis 2 (mechanical engine field test cell). ?

Situation Awareness of Engine Test Cell-Qualitative Reasoning

The objective here is to access the qualitative performance of a restricted engine test cell for an underground mine. The underlying objectives are to assess (i) temperature and get a feel for its flow, (ii) CO_2 and NO_2buildup, and (iii) other possible fumes.

Decision-Making: Sample Qualitative Reasoning at Time Step 80.

? Q_1What is happening?

? A_1: From $\boxtimes_1$–$\boxtimes_5$, at time step 80, NO_2 reduces from 0.5 to 0.08, CO_2 from 400 to 300 ppm, though the engine emitsmoreCO_2 in $\boxtimes_2$;H_2 fumeswere detected in the cell around time step 20, but disappeared over time with CH_4; temperature gradually increases across $\boxtimes_1$–$\boxtimes_5$.

? Q_2: *Why is it happening?*

? A_2: The air extractor proves effective, as the concentration drops towards the outlet $\boxtimes_5$; an eventwith fire took place at step 20, giving rise to H_2 fumes.

? Q_3: *What can we do about it?*

? A_3:The air extractormust bemaintained and declared effectively before testing any engine.

? Q_4: *What will happen next?*

? A_4: Staying away from $\boxtimes_2$ is recommended in view of too high CO_2 concentration, though the entire test cell seems safe to engineers.

Situation Awareness of Engine Test Cell-Quantitative Reasoning

From the real-time monitoring of the engine test cell, we specifically assess the quantitative performance of our ambient intelligence approach to consolidate the qualitative reasoning ranging from the use of temporal statistics to finding dependencies between parameters. We set and implemented five research questions in the following. These assist in sending early warning whenever hazardous observations are made over time, as indicated in Figure 6. We want to know the following over time.

?q_1: What is the level of health safety in each of the three regions in the engine test cell?

?q_2: What concentrations of the gases are commonly sensed in most regions?

?q_3: At what temperature levels are the gases' concentrations being sensed?

?q_4: Are there any interesting events that could be picked up from the gases' distributions??q_5: Is there any relationship between CH_4??concentrations and the temperature of the region?

Having implemented the temporal statistical models, the results obtained at time step 80 in Tables 2–4 provide answers to the research questions and substantiate the qualitative results reflected in Figure 9. It can be noted that all observed results are quite accurate from a general standpoint, as most standard errors compared to the mean values are less than 5% in all regions.

Table 3: Quantitative results of region 1 shown in Figure 9(a)

0 80	Temp	CO2	NO2	CH4	H2
Mean	17.02889	369.2593	0.208272	1.301235	0.223457
Standard error	0.09528	24.97194	0.017473	0.299694	0.097962
Median	17	360	0.18	0	0
Mode	16	50	0.04	0	0
Standard deviation	0.857524	224.7475	0.157256	2.697244	0.881656
Skewness	−0.09152	0.917227	1.269842	1.814768	3.777197
2= 397.37		df = 340		Prob. = 0.0173	

Table 4: Quantitative results of region 2 shown in Figure 9(b)

0 80	Temp	CO2	NO2	CH4	H2
Mean	17.50074	612.9259	0.080642	1.230864	1.033333
Standard error	0.077014	32.35313	0.003432	0.194949	0.212597
Median	17	581	0.08	0.1	0
Mode	17	999	0.04	0	0
Standard deviation	0.693128	291.1782	0.03089	1.75454	1.913374
Skewness	0.985312	0.023658	0.225995	1.176821	2.465437
2= 696.99		df = 312		Prob. = 2.00 – 31	

An interesting observation that can be made from the engine test cell is that the air extractor inlet ($\boxtimes_1$) and outlet ($\boxtimes_5$) are working perfectly. The reason for this is that the average of most of the gases suspended in air is more in $\boxtimes_2$, but lower in regions $\boxtimes_1$ and $\boxtimes_5$. In Table 3, $\boxtimes_1$ is largely safe when the mean values are compared to the normal gases' values in Table 1. In Table 4, region 2 is where the engine is run; the system becomes aware of the emission of more CO_2, that is, increasing concentration to 900 ppm, making it unsafe, unlike other gases. For the outlet in $\boxtimes_5$, however, Table 5 shows a quick reduction in CO_2 to 300 ppm, making the region safe, as it is for other gases. This answers question q_1 and autonomously sends an sms to safety agents.

Table 5: Quantitative results of region 5 shown in Figure 9(c)

0 80	Temp	CO2	NO2	CH4	H2
Mean	20.17654	359.8519	0.06537	2.447654	1.174074
Standard error	0.214816	15.80476	0.011157	0.285151	0.214534
Median	20.9	340	0.046	2.2	0
Mode	21	299	0.04	0	0
Standard deviation	1.933344	142.2428	0.100415	2.566361	1.930806
Skewness	−0.27947	0.982282	8.174952	0.706862	1.569931
2= 802.87	df = 875			Prob. = 0.9606	

At the 80th time step, concentrations commonly sensed in regions ⊠$_1$, ⊠$_2$, and ⊠$_5$ are shown on the modal rows, within temperature = 16°C, 17°C, and 21°C, respectively. The temperature reveals the direction of heat flow. The tables show that CH_4 and H_2 are rarely found in the cell, while NO_2 is constantly low. It is concluded that enough noise is present at the inlet in region 1 with a low 50? ppm concentration of CO_2 compared with other regions. This answers questions q_2 and q_3 and autonomously sends a message to the mobile phones of the safety agents.

The values of skewness in the three tables are greater or less than 0, showing clearly that events took place in the engine test cell: (i) three safety agents opened a door into/out of the cell, and (ii) a fire was ignited from a nozzle for tracing wind direction in the cell, leading to the detection of H_2 gas in the process. This actually correlates with the results in Figure 9 and answers question q_4.

Since the degrees of independence or probability (of ⊠^{2} in ⊠$_1$ and ⊠$_2$) < 0.05, it implies that there is a relationship between CH_4 and temperature. However, there may not be a relationship between them in region 3, since the probability (= 0.9606) > 0.05. It reveals that a higher temperature reflects a high CH_4 concentration and vice versa. This interestingly conforms to the properties of CH_4 shown in Table 1. This answers question q_5.

There are diverse discoveries of knowledge that the system could reveal, but our main questions q_n are needed in order to focus on the hypotheses. Hypothesis 2 states that the performance of remote monitoring of gases in restricted areas, such as the engine test cell, over WSNs is not correlated to its performance in areas such as those typically found in underground mines. These experiments, however, showed that this is not the case, implying that they are correlated.

Benchmarking Our Quantitative Reasoning with Publicly Available Observations and Methods

One of the ways to substantiate the reliability of our demonstrator is to benchmark its quantitative reasoning with publicly available air observations and methods. The Texas commission on environmental air quality captures hourly data collected, in particular, in the CPS Pecan

Valley C678, Heritage Middle School C622, and Calaveras Lake C59 regions of San Antonio metropolitan area, USA, in [29]. The information is updated hourly, and it is officially certified by the technical staff. The air gas pollutants considered in this study are CO, SO_2, and NO_2, with different temperatures in an apparent one-hour time lag in the data. The sample data for August 25, 2012, was integrated into our temporal statistical models implemented in a Python distributed programming environment.

The temporal quantitative results obtained at time step 23?pm in Table 6 provide answers to most of the research questions, especially question q_1 on the level of health safety in the metropolitan area, which the Texas commission provides for benchmarking. An interesting observation that can be made from the area is that the average of most of the gases suspended in air is fairly higher in the Heritage Middle School C622 and Calaveras Lake C59 regions, but lower in the CPS Pecan Valley C678 region. In Table 6, our temporal quantitative reasoning shows that the top region (CPS Pecan) is fairly safe when the mean values of the gases are compared to the acceptable values in air, except for NO_2, which is 1.896?ppm. In the middle region in Table 6, the mean values of SO_2 and NO_2 gases increase to concentrations of 2.608?ppm and 1.658?ppm, respectively, making it manageable or fairly unsafe, except CO, which is within the acceptable range. For the bottom region, however, Table 6 shows relatively more NO_2 with 1.642?ppm, making the region fairly unsafe/manageable, though other gases are not recorded. Although our early warning decision to the safety agents is very meticulous, one can see that our quantitative reasoning on the level of health safety is well correlated with the Texas air quality index (AQI) decisions. This answers the main question q_1, which benchmarks the two methods and shows the reliability of our approach.

Table 6: Benchmarking our approach with USA San Antonio metropolitan regions and its AQI method determining the health safety status

Regions	12:00 23:00	Temp	CO	SO2	NO2	Our approach	Texas AQI
	Mean	82.820833	0.291667	0.020833	1.895833		
	Standard error	1.185525	0.005763	0.020833	0.129726		
	Median	82.05	0.3	0	1.95		
CPS Pecan Valley c678	Mode	77	0.3	0	2	Fairly safe	Good
	Standard deviation	5.807864	0.028233	0.102062	0.635527		
	Kurtosis	−1.002790	9.123966	24	−0.701763		
	Skewness	0.368065	−3.219960	4.898979	0.265184		
	Mean	83.545833	0.4	2.608333	1.658333		
	Standard error	1.112332	2.31498 – 17	0.859725	0.441297		
	Median	82.65	0.4	0.2	0.6		
Heritage Middle School	Mode	77.1	0.4	0	0	Fairly unsafe	Moderate
	Standard deviation	5.449290	1.1341 – 16	4.211776	2.161907		
	Kurtosis	−0.947188	−2.190476	3.683374	2.008987		

	Skewness	0.473571	−1.067940	1.889303	1.575467		
	Mean	84.35	0	0	1.641667		
	Standard error	1.448200	0	0	0.124516		
	Median	83.5	0	0	1.45		
Calaveras Lake C59	Mode	78.5	0	0	1.5	Fairly unsafe	Moderate
	Standard deviation	7.094701	0	0	0.610001		
	Kurtosis	−0.903842	0	0	1.366540		
	Skewness	0.464159	0	0	1.440687		

Comparative Evaluations of the Proposed Framework with Other Related Methods

The evaluation of the contribution to mine safety using our proposed framework indicates that it is similar to the contributions derived from the related methods. A comparison of the related methods and the proposed framework is summarised in Table 7. The comparison and application of the related methods and the proposed framework to achieve mine safety and monitoring of toxic gases in particular show that our approach is comparatively complementary and building on the related baseline methods with regard to the following aspects.

- Methodology: (i) the proposed system is a reactive remote monitoring framework developed for real-time gas sensing, which spans physical layers through application space; (ii) the modelling of the theory in WSNs, Ohm's law, concepts of static and mobile robot sensing, and derivations of curvilinear gas calibration equations are intended to achieve autonomy; (iii) our approach implements distributed sensors more cheaply than the big sophisticated sensors, which could be laborious to set up or expensive for developing mines/countries. Thus, every methodology in Table 7 is obviously driven by the problem focus.
- Decision support system: (i) integrating intelligent DSS with distributed WSN for remotely monitoring toxic gases builds on the baseline produced by the related methods; (ii) the ambient intelligence, based on temporal statistical methods governing decision-making for miners, works well with both qualitative as well as quantitative parameters; (iii) this decision-making technique is comparatively easy to use, as it is lighter than using computationally time-consuming algorithms that could be heavy weight, affecting our real-time system with time lag.
- Field work: (i) our approach monitors gases remotely in a real-life engine test cell, such as those typically found in underground mines, generating knowledge to improve safety with an assurance of assistance whenever required; (ii) knowledge of how to implement DSS and WSN approaches and how to make them profitable in improving miners' safety is demonstrated; (iii) the proposed model not only assists mine management/companies in doing simulations like most related methods, but has also

done real-life field work indoors and in an aeronautic engine test cell, similar to those typically found in underground mines.

Table 7: Comparative evaluations of mine safety methods for toxic gases

Related work	Problem focus/ objective	Methodology	Results	DSS	Field work
Minming et al. 2009 [17]	Data aggregation modelling	(i) Directed diffusion (ii) Geographical information metrics	Addresses: data redundancy and transmission delay	N/A	(i) N/A (ii) Simulation
Yuan et al. 2009 [18]	High energy consumption in long-distance roadway	Uneven fixed cluster and mixed routing	Good network lifetime and connectivity	N/A	(i) N/A (ii) Simulation
Niu et al. 2007 [19]	Design WSN for mine safety monitoring	Overhearing-based adaptive data collecting scheme	Reduces traffic and control overheads	N/A	Real-life test
Shaoet al. 2008 [20]	Safety monitoring system of mine	Ad hoc technology	Design of physical, MAC, and network layers	Manual	Not specified
Wu et al. 2010 [21]	Network deployment and multihop routing patterns	BRIT (bounce routing in tunnels) algorithm	Hybrid signal propagation model in three-dimensional underground tunnels	N/A	(i) N/A (ii) Simulation
Government practices [22]	Monitoring toxic gases	(a) Trainings (b) Ventilation (c) Lag time (d) Big sensors	Laborious and inaccurate	Manual	Reallife

Proposed system model	Decision-making in monitoring toxic gases	(i) Distributed portable sensors (ii) Calibration algorithm (iii) Regression models (iv) Ohm's law (v) Static and mobile sensing and phones	(i) Satisfactory (ii) Real-time and remote monitoring	(i) Autonomous (ii) Ambient intelligence (iii) Temporal statistical models	(i) Real-life test (ii) Aeronautic engine test cell (iii) Indoors

CONCLUDING REMARKS

The study on remote monitoring of toxic gases targeting underground mines has analysed various aspects of using a WSN, SA, and temporal statistical models as ambient intelligence in an attempt to effect autonomous decision-making in real time. Some of the main conclusions that can be drawn from the experiments and analysis are as follows.

- Every implementation node in our framework functions as well as the pioneer robot when all its parameters are configured appropriately.
- The experiments confirmed some of the claims discovered in the literature survey about toxic gases, Ohm's law, statistical methods, the theory of SA, and direct correlation of the properties of the engine test cell and underground mines.
- Even the simple ambient intelligence proposed, based on the temporal statistics, provides better results than common qualitative decision-making through consolidation.
- The chosen gas calibration equations may have considerable effects on the accuracy of the ppm results.
- It is assumed that the robot navigates moderately to add extra airflow to the sensing node as it perceives higher evident peaks.

While working on the research project, a number of possible variations on the experiments were considered, yet not implemented. These are listed here as topics for future research:

- detecting redundant sensor nodes and optimal sensor placement

- autonomous determination of battery lifetime and remotely hibernating battery power
- noise removal and alternative decision-making models
- integrating Figures 2 and 3 to improve safety in underground mines.The approach of defining hypotheses before performing experiments has advanced researchers' knowledge in an intriguing approach to studying problems and their potential solutions. Thus, this is a directed research project affecting lives by improving health safety and therefore ready for deployment in underground mines.

ACKNOWLEDGMENTS

The author gratefully acknowledges resources and financial support made available by the University of South Africa (UNISA) and the Council for Scientific and Industrial Research (CSIR), South Africa.

REFERENCES

1. M. Möhner, M. Lindtner, H. Otten, and H. G. Gille, "Leukemia and exposure to ionizing radiation among German uranium miners," American Journal of Industrial Medicine, vol. 49, no. 4, pp. 238–248, 2006.
2. M. C. Betournary, "Underground mining and its surface effects," Interstate Technical Group on Abandoned Underground Mines Fourth Biennial Abandoned Underground Mine Workshop, Report MMSL 02-021, 2010.
3. M. Kreuzer, L. Walsh, M. Schnelzer, A. Tschense, and B. Grosche, "Radon and risk of extrapulmonary cancers: results of the German uranium miners' cohort study, 1960–2003," British Journal of Cancer, vol. 99, no. 11, pp. 1946–1953, 2008. · ·
4. MHSC, http://www.mhsc.org.za/.
5. R. S. Conti, Fire-fighting resources and fire preparedness for underground coal mines, National Institute for Occupational Safety and Health Pittsburgh Research Laboratory, 1994.

6. Swansea Station, http://www.swansea.airqualitydata.com/cgi-bin/faqs.cgi?1009.
7. Ohio Station, http://www.nytimes.com/2006/09/27/us/27steubenville.html.
8. E. R. Musaloiu, A. Terzis, K. Szlavecz, A. Szalay, J. Cogan, and J. Gray, "Life under your feet: a wireless soil ecology sensor network," in Proceedings of the 3rd IEEE Conference on Embedded Networked Sensors (EmNetS ‹06), 2006.
9. S. Li, S. H. Son, and J. A. Stankovic, "Event detection services using data service middleware in distributed sensor networks," in Proceedings of the 2nd International Conference on Information Processing in Sensor Networks (IPSN ‹03), pp. 502–517, Springer, Berlin, Heidelberg, 2003.
10. A. B. Bagula, I. Osunmakinde, and M. Zennaro, "On the relevance of using Bayesian belief networks in wireless sensor networks situation recognition," Sensors, vol. 10, no. 12, pp. 11001–11020, 2010. · ·
11. T. Melodia, M. C. Vuran, and D. Pompili, "The state-of-the-art in cross-layer design for wireless sensor networks," in Wireless System/Network Architect, M. Cesana and L. Fratta, Eds., vol. 3883 of Lecture Notes in Computer Science, pp. 78–92, Springer, Heidelberg, Germany, 2006.
12. M. G. D. Benedetto and G. Giancola, Understanding Ultra Wideband Radio Fundamentals, Prentice Hall, Upper Saddle River, NJ, USA, 2004.
13. M. Z. Win and R. A. Scholtz, "Ultra-wide bandwidth time-hopping spread-spectrum impulse radio for wireless multiple-access communications," IEEE Transactions on Communications, vol. 48, no. 4, pp. 679–691, 2000.
14. Y. Yuan, Z. Shen, W. Quan-Fu, and S. Pei, "Long distance wireless sensor networks applied in coal mine," Journal Procedia Earth and Planetary Science, vol. 1, pp. 1461–1467, 2009.
15. K. R. Rogers, S. L. Harper, and G. Robertson, "Screening for toxic industrial chemicals using semipermeable membrane devices with rapid toxicity assays," Analytica Chimica Acta, vol. 543, no. 1-2, pp. 229–235, 2005. · ·

16. Inspectapedia, http://inspectapedia.com/hazmat/Environment_Building.php.
17. T. Minming, N. Jieru, W. Hu, and L. Xiaowen, "A data aggregation model for underground wireless sensor network," in Proceedings of the WRI World Congress on Computer Science and Information Engineering (CSIE ‹09), vol. 1, pp. 344–348, April 2009. · ·
18. Y. Yuan, Z. Shen, W. Quan-Fu, and S. Pei, "Long distance wireless sensor networks applied in coal mine," Journal of Procedia Earth and Planetary Science, vol. 1, no. 1, pp. 1461–1467, 2009.
19. X. Niu, X. Huang, Z. Zhao, Y. Zhang, C. Huang, and L. Cui, "The design and evaluation of a wireless sensor network for mine safety monitoring," in Proceedings of the 50th Annual IEEE Global Telecommunications Conference (GLOBECOM ‹07), pp. 1291–1295, November 2007. · ·
20. Q. L. Shao, Z. X. Zhang, and Z. C. Zhang, "Research on safety monitoring system of mine based on ad hoc technology," in Proceedings of the International Conference on Information Management, Innovation Management and Industrial Engineering (ICIII ‹08), vol. 2, pp. 450–453, December 2008. · ·
21. D. Wu, L. Bao, and R. Li, "A holistic approach to wireless sensor network routing in underground tunnel environments," Journal of Computer Communications, vol. 33, no. 13, pp. 1566–1573, 2010.
22. Centers for Disease Control and Prevention, Atlanta, USA,http://www.cdc.gov/niosh/mining/topics/Ventilation.html.
23. R. Lin, Z. Wang, and Y. Sun, "Wireless sensor networks solutions for real time monitoring of nuclear power plant," in Proceedings of the 5th World Congress on Intelligent Control and Automation (WCICA ‹04), pp. 3663–3667, June 2004.
24. J. Hashem, T. Liu, Z. Liang, and J. Ye, "Using Ohm›s law to calibrate a picoammeter to 0.4 pico-ampere precision," American Journal of Undergraduate Research, vol. 10, no. 1, 9 pages, 2011.
25. http://www.libelium.com/products, http://www.futurlec.com/Gas_Sensors.shtml.
26. C. Sophocleous and R. J. Wiltshire, "On linearizing systems of diffusion equations," Journal of Symmetry, Integrability and Geometry, vol. 2, article 004, 11 pages, 2006.

27. M. R. Endsley, "Theoretical underpinnings of situation awareness: a critical review," in Situation Awareness Analysis and Measurement, M. R. Endsley and D. J. Garland, Eds., LEA, Mahwah, NJ, USA, 2000.
28. N. Maxwell, Data Matters: Conceptual Statistics for a Random World, Key College, Emeryville, Calif, USA, 2004.
29. Texas Commission on Environmental Air Pollution Data Quality, USA,http://www.tceq.texas.gov/airquality/monops/hourly_data.html.

Chapter 4

Introducing the Iron Potential Zones Using Remote Sensing Studies in South of Qom Province, Iran

Faranak Feizi and Edris Mansouri

Department of Mining Engineering, Faculty of Engineering, South Tehran Branch, Islamic Azad University, Tehran, Iran

ABSTRACT

The studied area—Kermejegan—is located in the south of Qom Province, Iran. In this paper, geology map, ASTER and ETM7+ satellite images were used and after processing these images with Geomatica and ENVI softwares, iron potential zones were compared with the iron mine position in the south of the area and 2 dominate indexs around. Finally remote sensing, faults and geological data layers were integrated in GIS and hopeful zones were introduced for continuing the exploration processes.

INTRODUCTION

The use of satellite images for mineral exploration has been very successful in pointing out the presence of minerals such as alunite, pyrophyllite, kaolinite, sericite, illite, muscovite, smectite, and carbonate which are important in the identification of hydrothermal alterations [1]. In addition, satellite remote sensing provides synoptic view, which is helpful in identification and delineation of various land forms, linear features and structural elements [2, 3]. Hence, in primary stages, mapping geologic lineaments are important as long-range management for mineral exploration, because of their potentials for harboring ore bodies that are carried and deposited by ascending hydrothermal fluids [4]. Iran is located in the Alpine-Himalian orogenic and metalogenic belt and has high potentials for gold and copper and other base metal deposits. Satellite images have been used by many Iranian geologist researchers as the cheapest method for interpretation of the structural features and exploration purposes [5, 6]. For instance, distinguished phyllic, propylitic and argillic alteration zones in the Urumieh-Dokhtar Volcanic Belt of Iran [7], separated a few hydrothermal alterations in the Valayesh in the northwestern Iran [6], distinguished phyllic and argillic alteration zones in the magmatic arc in southeast of Iran [8], separated hydroxyl and iron oxide minerals in the Yazd, central Iran [9], show that satellite images are useful for locating and interpretation of the structural features in the Zagros Structural Belt, in southwest of Iran [10]. The aims of the present study are to perform principal component analysis (PCA), apply spectral angle mapper (SAM), Least Square Fit (LS-Fit), Minimum Noise Fraction (MNF) and band ratio to map the iron zones on the ASTER (Advanced Space borne Thermal Emission and Reflection Radiometer) and ETM7+ satellite imagery data.

GEOLOGY SETTING

Kermejegan area is located between longitudes 481493 - 487103 and latitudes 3787390 - 3792893 in the south of Qom Province, Central Iran (Figure 1). The studied area is a small central part of 1:100000 Kahak geological map. The litho logy of this part includes volcanics such as Andesite, Andesite Basalt to mega porphyry Andesite. Different types of Sandi tuff and Limestone in south and southeast parts and quaternary units are seen in East parts of the area. Hematitization and limonitization are seen in north and southeast part [11]. There is an iron potential zone in southeast of the area with Magnetite and Hematite mineralization in vein and vein lets type which is in detail exploration phase now, showing in the map with green point. Also there are 2 iron mine in southwest of the area which are shown with red points in the map. As the map showing, the main fault—Bidhend—is going throw the south iron potential and central part of the area (Figure 2).

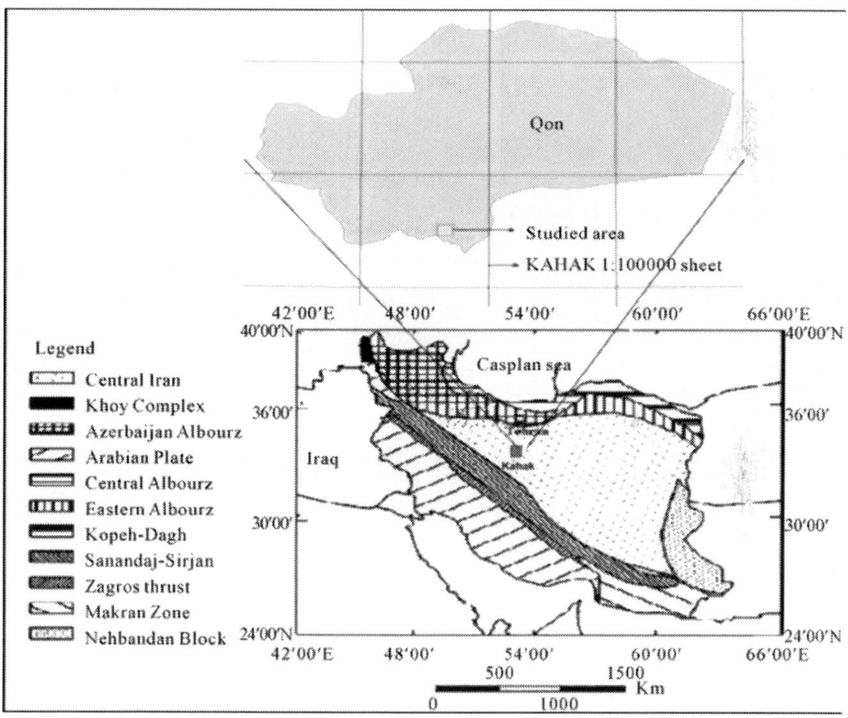

Figure 1: The studied area in the map of Iran.

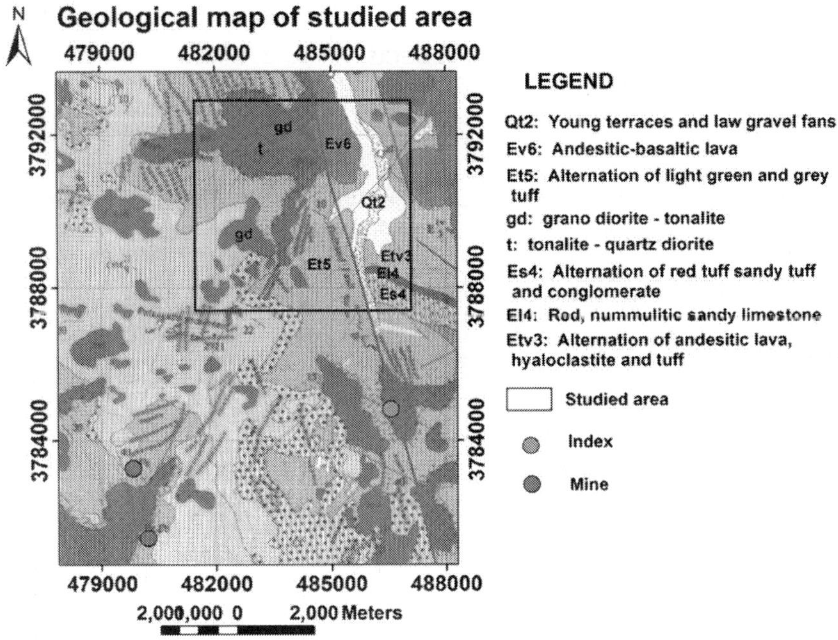

Figure 2: The geological map of studied area (1:100000 Kahak geology map).

MATERIALS AND METHODS

ETM7+ Data

Principal Components Analysis (PCA) method was used for doing this phase. After processing with Geomathica software, RGB: (5 - 7) & PC2 (5 - 7) & PC4 (1, 4, 5, 7) were used. The iron potential zones were shown in goldish yellow by this method (Figure 3).

ASTER Data

The ASTER is an advanced optical sensor comprised of 14 spectral channels ranging from the visible to thermal infrared region. It will provide scientific and also practical data regarding various field related to the study of the earth [12]. Various factors affect the signal measured at the sensor, such as drift of the sensor radiometric calibration, atmospheric and topographical effects. For accurate analysis, all of these corrections are necessary for remote sensing imagery. To this end, at the beginning of the path, data set in hierarchical data format (HDF) was used for this research and radiance correlation such as wavelength, dark subtract and log residual by ENVI4.4 software which is essential for multispectral images, were implemented. For ASTER image processing, MNF LS-Fit PCA SAM methods were used.

Integration of Data Layers in GIS

Data layers have to integrate in GIS for getting any results. For this reason the geological layer with the location of fault on that should be the main layer and then the iron oxide potential zones which recognized by remote sensing integrated with that. In next step iron mines and indexes integrated with them and at last the final map prepared.

RESULT AND DISCUSSION

Aster and ETM7+ Images Processing

Least Squares Fitting

Least Squares Fitting (LS-Fit) is the technique assumes that the bands used as input values are behaving as the variables of a linear expression and the "y" value of the equation, namely the predicted band information, gives us a calculated output value. This predicted band is what that band should be according to the linear equation. The minerals which are sensitive to a specific band are then differentiated from the features which are reflective to the other bands as well; just by taking the difference between the predicted values and the original values [13].

Distribution of iron oxide was created by using all the 3 visible and near-infrared (VNIR) bands as the input bands and VNIR-b1 as the modeled band (Figure 4).

Minimum Noise Fraction

The Minimum Noise Fraction (MNF) transformation is used to determine the inherent dimensionality of image data, segregate noise in the data, and reduce the computational requirements for subsequent processing [14-16]. MNF involves two steps: in first step which is also called noise whitening, principal components for noise covariance matrix are calculated. This step decorrelates and rescales the noise in the data. In second step principal components are derived from the noise whitened data. The data can then be divided into two parts. One part associated with large Eigen values and the other part with near unity Eigen values and noise dominated images. Using data with large Eigen values separates the noise from the data, and improves spectral results [15, 16]. MNF analysis can identify the locations of spectral signature anomalies. This process is of interest to exploration geologist because spectral anomalies are often indicative of alterations due to hydrothermal mineralization [16]. MNF band 2, was used for iron oxide alterations (Figure 5).

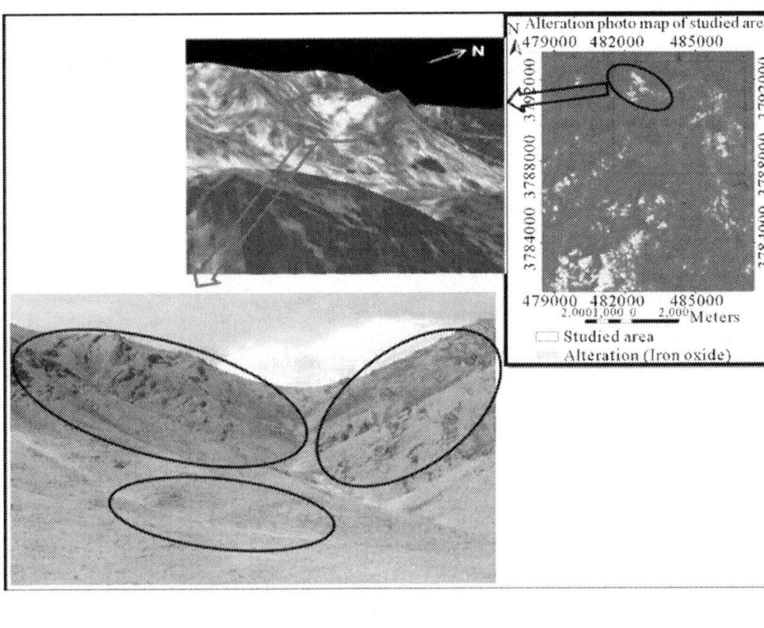

Figure 3: 2 and 3 dimension processing image with check field of iron oxide zones.

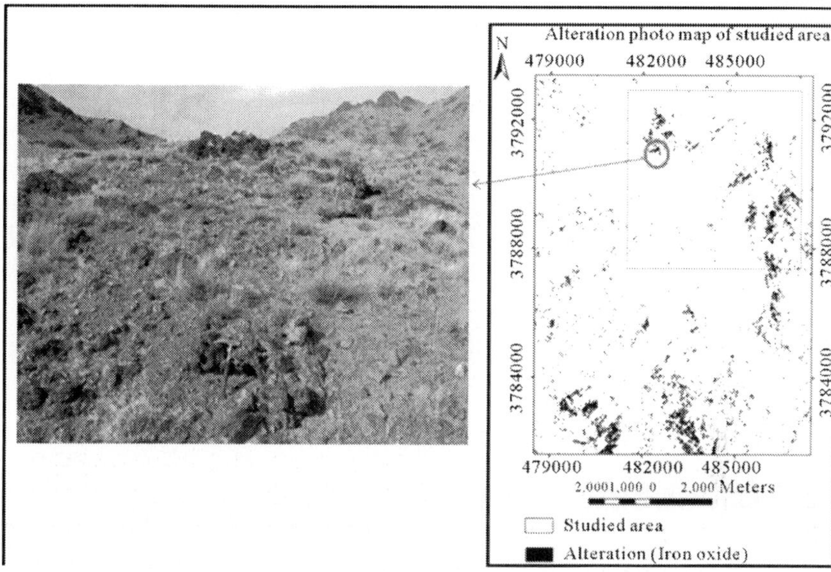

Figure 4: ASTER image processing with LS-Fit method and check field.

Principal Component Analysis

The Principal Component Analysis (PCA) is a multivariate statistical technique that selects uncorrelated linear combinations (eigenvector loadings) of variables. Each successively extracted linear combination, or principal component (PC), has a smaller variance. PCA is widely used for alteration mapping in metallogenic provinces [17]. PCA technique has been applied in this study. An approach based on the examination of eigenvector loadings in each PC image is used for determining which image contains information related to the spectral signatures of specific target minerals. It is expected that the PC image that collects moderate to high eigenvector loadings for the diagnostic absorptive and reflective bands of the index mineral could be considered as the specific image for that mineral. If the loading of absorptive band is negative in sign, the target area will be enhanced in bright pixels, and if the loading of reflective band is negative, the area will be enhanced in dark pixels [18, 19] (Figure 6).

Table 1 shows the eigenvector loadings for bands 1, 2, 3 and 4. Inverse of PC4 can show the areas with iron oxide.

Spectral Angle Mapper

The spectral angle mapper (SAM) is a classification technique that permits rapid mapping by calculating the spectral similarity between the image spectrums to reference reflectance spectra. The image spectra were compared with USGS Digital Spectral Library (Minerals) [20]. Figure 7showed selected minerals spectral library plots that related to iron oxide. 4 mineral spectral representative of iron oxide include jarosite, hematite, goethite, limonite was selected (Figure 7). SAM measures the spectral similarity by calculating the angle between the two spectra, treating them as vectors in n-dimensional space [16, 21] (Figure 8).

GIS Processing

In this part for integration the data layers in GIS area, first of all, the shape files of all alteration zones which were carried out with different methods were drown. Then the layers were overlapped on each other. Afterwards the most overlapped zones were chosen and were controlled with field investigations. As the last step, the geology map of studied area which was shown the iron potentials, was integrated with the final alteration map in GIS area. As it was shown in Figure 9, the iron potential zones (green squares) are similar to south and southeast mines and indexes in litho logy (Tuffs) and intrusive bodies. Also the processed spectrum gradation in those potential zones have good situations. So these two squares have been suggested for more exploration.

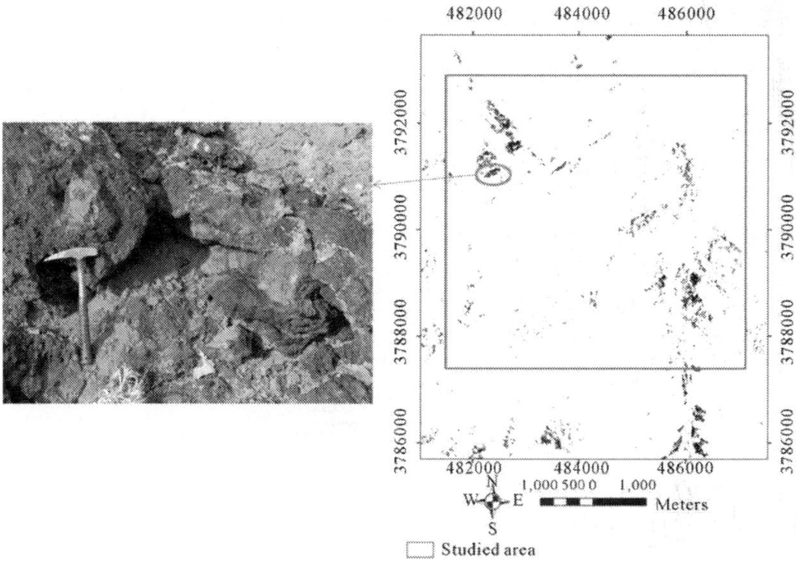

Figure 5: ASTER image processing with MNF method and check field.

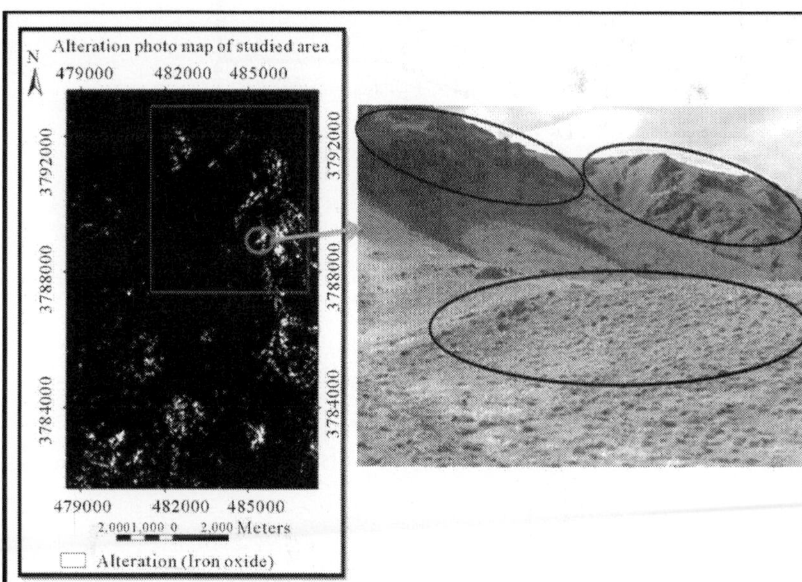

Figure 6: ETM7+ image processing with PCA method and check field.

Table 1: The result of PCA for enhancing iron oxide zone

Eigenvector	Band 1	Band 2	Band 3	Band 4
Band 1	0.392563	0.578218	0.553369	0.453146
Band 21	0.710413	0.296063	-0.418364	-0.482319
Band 3	0.005043	0.125901	-0.705904	0.697011
Band 4	0.584109	-0.749773	0.143019	0.276049

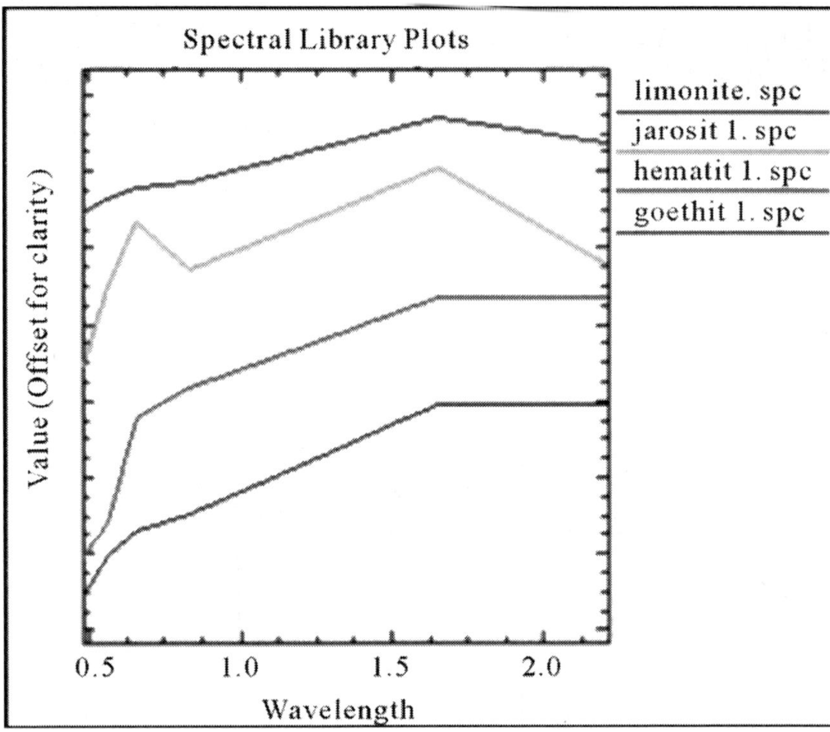

Figure 7: Spectral library plots from USGS library for iron oxide minerals.

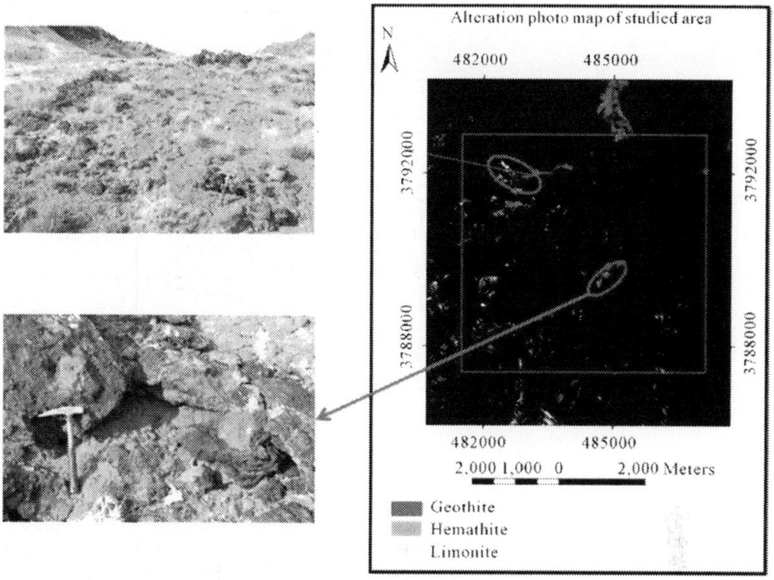

Figure 8: ASTER image processing with SAM method and check field.

Comparison between Iron Potential Zones

After all these software analyses, check field processes were necessary (Figures 3-6 and 8). The control points were detected, and after the check fields, the correction of alteration zones were confirmed. These figures show some check points for iron oxide which were recognized with remote sensing processes. These checks confirmed the results of the RS methods. As the last step the comparison between two suggested squares zones were necessary. So the Z. profiles of these two zones were prepared and compared with the Z. profiles of mine and indexes in south of the studied area. As it was shown in Figure 10, these profiles are so similar to each other. So the correction of these two suggested zones was confirmed again. Table 2 shows the results of Fe analyses on Sarvian mine in south of the studied area.

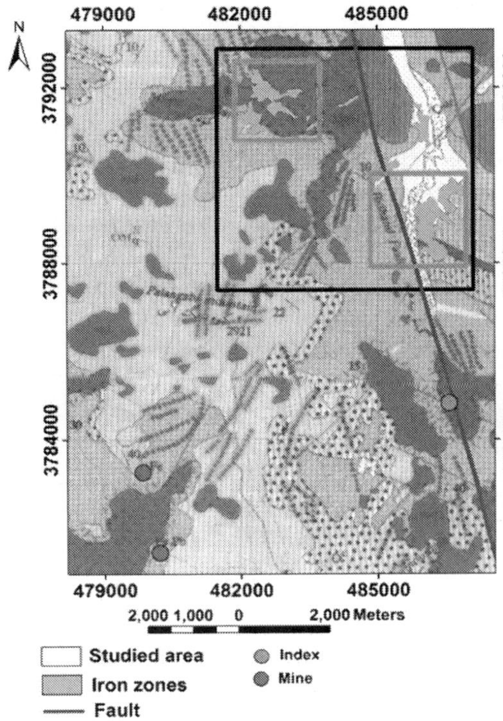

Figure 9: Final map, prepared by GIS.

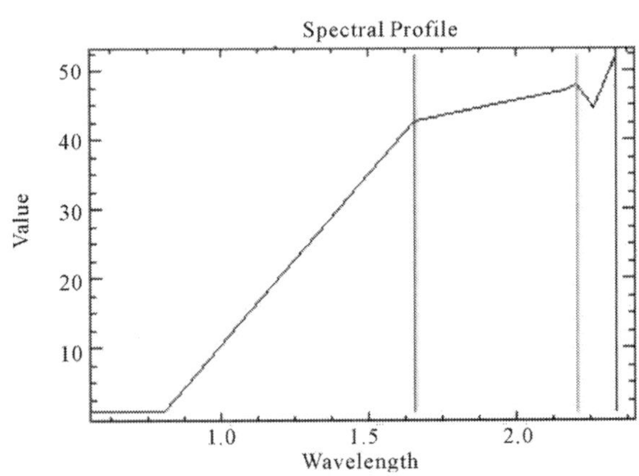

(a)

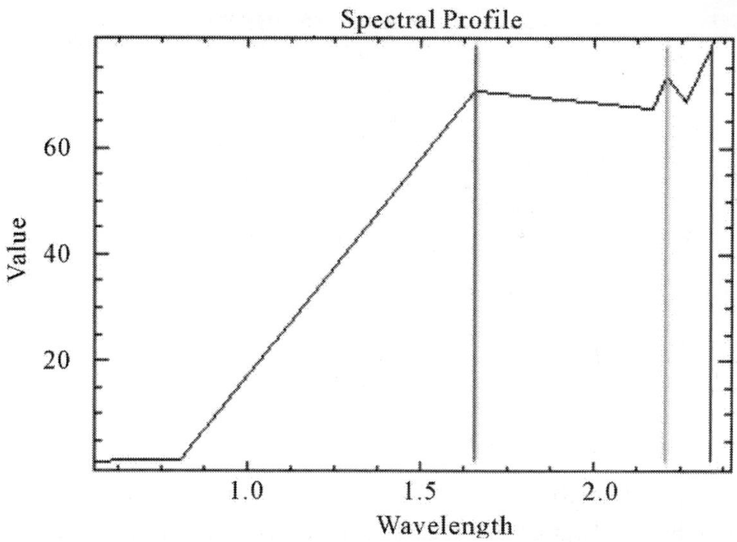
Spectral Profile
Value
60
40
20
1.0
1.5
2.0
Wavelength

(b)

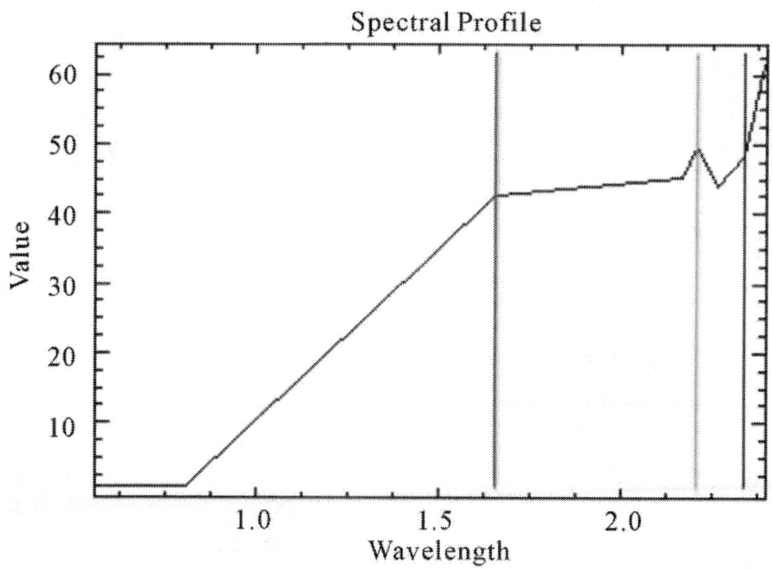
Spectral Profile
Value
60
50
40
30
20
10
1.0
1.5
2.0
Wavelength

(c)

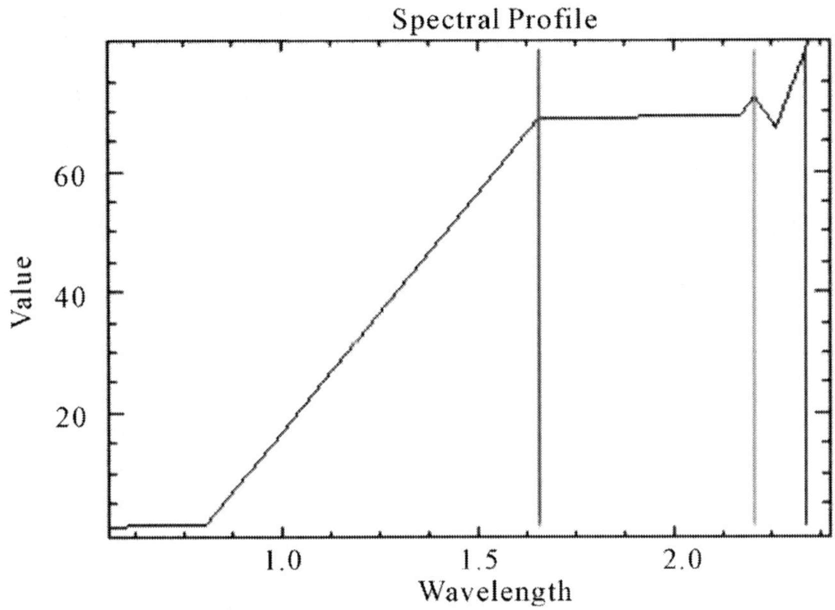

(d)

Figure 10: Z. profile of Fe Index (a), Z. profile of high potential zone in south east of studied area (b), Z. profile of Fe mine (c), Z. profile of high potential zone in North West of studied area (d).

Table 2: Results of Sarvian Fe mine analyses

Sample No.	Location		Fe% Total
	X	Y	
1	480352	3781509	38
2	480348	3781474	59
3	480356	3781422	46

CONCLUSIONS

There are two hopeful areas which suggested for more exploration. Because of the main fault of the studied area going to throw the iron mineralization zones, determination of lineaments in order to connection between structural elements and mineralization has suggested.

REFERENCES

1. S. M. Salem, S. A. Arafa, T. M. Ramadan and E. S. A. El Gammal, "Exploration of Copper Deposits in Wadi El Regeita Area, Southern Sinai, Egypt, with Contribution of Remote Sensing and Geophysical Data," Arabian Journal Geoscience, Vol. 6, No. 2, 2011, pp. 321-335. doi:10.1007/s12517-011-0346-z
2. A. Javed and M. H. Wani, "Delineation of Groundwater Potential Zones in Kahund Watershed, Eastern Rajasthan, Using Remote Sensing and GIS Techniques," Journal Geological Society of India, Vol. 73, No. 2, 2009, pp. 229- 236. doi:10.1007/s12594-009-0079-8
3. M. I. Adham, C. S. Jahan, Q. H. Mazumder, M. M. A. Hossain and A. L. Haque, "Study on Groundwater Recharge Potentiality of Barind Tract, Rajshahi District, Bangladesh Using GIS and Remote Sensing Technique". Journal Geological Society of India, Vol. 75, No. 2, 2010, pp. 432-438. doi:10.1007/s12594-010-0039-3
4. C. P. Kujjo, "Application of Remote Sensing for Gold Exploration in the Nuba Montains, Sudan," Master of Science Thesis, Bowling Green State University, Wooster, 2010.
5. H. Asadi Haroni and A. Lavafan, "Integrated Analysis of ASTER and Landsat ETM Data to Map Exploration Targets in the Muteh Gold-Mining Area, IRAN," 5th International Symposium on Spatial Data Quality, Enschede, 2007.
6. H. Azizi, M. A. Tarverdi and A. Akbarpour, "Extraction of Hydrothermal Alterations from ASTER SWIR Data from East Zanjan, Northern Iran," Advances in Space Research, Vol. 46, No. 1, 2010, pp. 99-109. doi:10.1016/j.asr.2010.03.014

7. A. B. Pour and M. Hashim, "Identifying Areas of High Economic-Potential Copper Mineralization Using ASTER Data in the Urumieh-Dokhtar Volcanic Belt, Iran," Advances in Space Research, Vol. 49, No. 4, 2012, pp. 753-769. doi:10.1016/j.asr.2011.11.028
8. J. C. Mars and L. C. Rowan, "Radiometer (ASTER) Data and Logical Operator Algorithms Arc, Iran, Using Advanced Spaceborne Thermal Emission and Reflection Regional Mapping of Phyllic and Argillic Altered Rocks in the Zagros Magmatic," Geosphere, Vol. 2, No. 3, 2006, pp. 161-186. doi:10.1130/GES00044.1
9. R. Poormirzaee and M. M. Oskouei, "Use of Spectral Analysis for Detection of Alterations in ETM Data, Yazd, Iran," Applied Geomatics, Vol. 2, No. 4, 2010, pp. 147- 154.doi:10.1007/s12518-010-0027-8
10. S. A. Ali and S. Pirasteh, "Geological Applications of Landsat Enhanced Thematic Mapper (ETM) Data and Geographic Information System (GIS): Mapping and Structural Interpretation in South-West Iran, Zagros Structural Belt," International Journal of Remote Sensing, Vol. 25, No. 21, 2004, 4715-4727.doi:10.1080/01431160410001688295
11. A. Aghanabati, "Geology of Iran," Geological Survey of Iran, 2004, 622 p.
12. H. Watanabe and K. Matsuo, "Rock type Classification by Multi-Band TIR of ASTER," Geoscience Journal, Vol. 7, No. 4, 2003, pp. 347-358. doi:10.1007/BF02919567
13. E. V. Yetkin, Toprak and M. L. Suezen, "Alteration Mapping By Remote Sensing: Application to Hasanda — Melendiz Volcanic, Complex," Geo-Imagery Bridging Continents 10th ISPRS Congress, Istanbul, 2004.
14. J. W. Boardman, F. A. Kruse and R. O. Green, "Mapping Target Signatures via Partial un Mixing of AVIRIS Data," Summaries, Proceedings of the 5th JPL Airborne Earth Science Workshop, Pasadena, 1995, pp. 23-26
15. A. A. Green, M. Berman, P. Switzer and M. D. Craig, "A Transformation for Ordering Multispectral Data in Terms of Image Quality with Implications for Noise Removal," IEEE Transactions on Geoscience and Remote Sensing, Vol. 26, No. 1, 1988, pp. 65-74.doi:10.1109/36.3001

16. A. B. Pour, M. Hashim and M. Marghany, "Using Spectral Mapping Techniques on Short Wave Infrared Bands of ASTER Remote Sensing Data for Alteration Mineral Mapping in SE Iran," International Journal of the Physical Sciences, Vol. 6, No. 4, 2011, pp. 917-929.
17. S. Myint, K. Aung and T. Isao, "Application of Remote Sensing Techniques on Iron Oxide Detection from ASTER and Landsat Images of Tanintharyi Costal Area, Myanmar," Scientific and Technical Reports of Faculty of Engineering and Resource Science No. 26, Akita University, Akita, 2005, pp. 21-28.
18. A. P. Crosta and J. M. Moore, "Enhancement of Landsat Thematic Mapper Imagery for Residual Soil Mapping in SW Minais Gerais State, Brazil: A Prospecting Case History in Greenstone Belt Terrain," Proceedings of the 7th Thematic Conference on Remote Sensing for Exploration Geology, 1989.
19. M. Khaleghi and H. Ranjbar, "Alteration Mapping for Exploration of Porphyry Copper Mineralization in the Sarduiyeh Area, Kerman Province, Iran, Using ASTER SWIR Data," Australian Journal of Basic and Applied Sciences, Vol. 5, No. 8, 2011, pp. 61-69.
20. A. Malekzadeh, M. H. Karimpour, C. R. Stern and S. A. Mazaheri, "Hydrothermal Alteration Mapping in SW Birjand, Iran, Using the Advanced Spaceborne Thermal Emission and Reflection Radiometer (ASTER) Image processing," Journal of Applied Sciences, Vol. 9, No. 5, 2009, pp. 829-842.
21. F. A. Kruse, J. W. Boardman, A. B. Lefkoff, K. B. Heidebrecht, A. T. Shapiro, P. J. Barloon and A. F. H. Goetz, "The Spectral Image Processing System (SIPS)—Interactive Visualization and Analysis of Imaging Spectrometer Data," Remote Sensing of Environment, Vol. 44, No. 2-3, 1993, pp. 145-163.

Chapter 5

Web of Things-Based Remote Monitoring System for Coal Mine Safety Using Wireless Sensor Network

Cheng Bo[1], Cheng Xin[2], Zhai Zhongyi[1], Zhang Chengwen[3], and Chen Junliang[1]

[1]State Key Laboratory of Networking and Switching Technology, Beijing University of Posts and Telecommunications, Beijing 100876, China

[2]Sichuan Province Key Laboratory of Geological Nuclear Technology, Chengdu University of Technology, Sichuan 610059, China

[3]Beijing Key Laboratory of Intelligent Telecommunications Software and Multimedia, Beijing University of Posts and Telecommunications, Beijing 100876, China

ABSTRACT

Frequent accidents have occurred in coal mine enterprises; therefore, raising the technological level of coal mine safety monitoring systems is an urgent problem. Wireless sensor networks (WSN), as a new field of research, have broad application prospects. This paper proposes a Web of Things- (WoT-) based remote monitoring system that takes full advantage of wireless sensor networks in combination with the CAN bus communication technique that abstracts the underground sensor data and capabilities into WoT resources to offer services using representational state transfer (REST) style. We also present three different implemented scenarios for WoT-based remote monitoring systems for coal mine safety, for which the system performance has been measured and analyzed. Finally, we describe our conclusions and future work.

INTRODUCTION

The natural conditions of coal mines are highly complicated, and mining conditions are extremely capricious. Many disasters can occur in mines, which increases the insecurity of coal mining and easily leads to major accidents, causing extreme difficulty in establishing safety. The structure of a coal mine environment is complex; the space for branch tunnels is limited, and the directions of branch tunnels are not fixed. Wired transmission systems are often installed only in the main tunnel, which substantially limits the expansion of the network. When underground mining advances continuously, no wired network can be established in real time; naturally, it is thus impossible to monitor these dangerous regions in real time. In addition, due to cost and maintenance limitations, no safety monitoring systems are installed in abandoned underground tunnels, creating a great potential safety hazard. Thus, for coal mine safety monitoring and control, there are still many shortcomings in wired monitoring and control that should be addressed. It is not easy to install wired coal mine safety monitoring and control systems in many coal mine regions, such as abandoned tunnels and mining sections. However, it is in these regions where there is the greatest potential safety hazard. Catastrophic coal mine accidents may occur if close attention is not paid, with inestimable consequences. Therefore, wired monitoring and control systems are not sufficient to meet the requirements of the safety production monitoring of an entire coal mine. Due to the limitations of wired monitoring and control systems and the complexity of underground coal mines, certain underground parameters, such as methane, dust, negative pressure, temperature, carbon monoxide, and wind speed, cannot be effectively monitored and controlled, causing great potential safety hazards [1–6]. Combining wireless monitoring systems with the existing wired monitoring and control systems can effectively compensate for the shortcomings of the existing monitoring and control systems, significantly increasing the level of coal mine safety.

For coal mine monitoring systems, data communication technology is the key to achieving the intellectualization and network of such systems and is also an important link to achieving a high-speed information platform for coal mines. Currently, the common communication modes that can be operated in the coal mine industry include the modulation and demodulation-based analog transmission mode, the distributed control system- (DCS-) based RS485 serial communication mode, the fieldbus control system (FCS) structure-based intelligent sensor transmission mode, and the wireless transmission mode. The fieldbus technique has been widely used in the industrial fields of various industries and can be used in either the master-slave mode or the multimaster mode as needed. The multimaster mode allows data exchanges among monitoring devices in different locations to be more flexible and direct. Due to its extremely high reliability and unique design, as well as its transmission characteristics (high speed, long distance), the fieldbus technique is especially suitable for use in the interconnection between field monitoring devices.

Established by densely distributing a large number of microsensor nodes capable of communicating and calculating in the unattended monitor regions, wireless sensor networks (WSNs) are intelligent autonomous control network systems that can self-complete planned tasks according to the environment. Based on the aforementioned main issues in the existing coal mine gas monitoring systems, the technological advantages of WSNs can be used to achieve seamless underground safety monitoring coverage by combining their advantages with those of wired networks.

Web services [7–9] are the services provided via Web technology. Generally, Web services are based on hypertext transfer protocol (HTTP). The World Wide Web Consortium (W3C) defines Web services as software systems marked by uniform resource identifiers (URIs); the common Web interface is defined and described by extensible markup language (XML) documents to provide outer calling. During the calling process, a Web service interacts with another Web service or other software systems using XML information encapsulation via HTTP. The Web is often used as the message transfer format between different services.

Representational State Transfer (REST) [10–13] is a network application framework first proposed by Dr. Roy Fielding in his doctoral dissertation. In the REST framework, all objects and capabilities within a system can be abstracted as resources, and every resource has a uniform resource location (URL). For instance, one data entry, one service operation, one picture, and one video can each be abstracted as a resource, and each resource has a unique URL within the system. In the REST framework style, every resource is exposed to the clients via a unique URL. Therefore, clients can distinguish these interactive objects via their URI. In addition, clients can also make any request to the resource based on a certain URI. One major difference between REST style Web services and traditional Web services is that REST does not treat HTTP as the transport layer protocol but the application layer protocol. Because the URI of a resource is unique, when a client uses the HTTP standard to operate GET, POST, PUT, and DELETE to request the target resource, the server will understand the HTTP request. GET reads the data of the resource; POST establishes a new resource; PUT updates a resource; and DELETE deletes a resource. Operating resources via a uniform interface increases the unity of interaction, reduces the coupling degree between components, simplifies the realization of the client, and increases the extensibility of the system.

The Web of Things (WoT) [14, 15] describes the Internet of things from the aspect of technical implementation. The WoT uses the design concept and technology of the Web to abstract the devices in the network environment of the WoT into resources and service abilities and connects them to the Web space to establish a heterogeneous network and distributed terminal-based application environment, which allows easier access to the embedded devices and services on the WoT. The WoT uses the Web standard and expands the entire ecosystem of the Internet to all types of sensing devices.

In this study, a wireless sensor network was combined with the controller area network (CAN) bus technology for the comprehensive and timely monitoring and intelligent early warning in the underground environment, the production data, and the operating state of the equipment. In addition, based on the WoT technology, all types of parameters were collected and transmitted to the remote monitor center for analysis to provide decision-making information for clients. If there was a parameter anomaly, the local sensor node could set off the sound and light alarm on site and simultaneously set off the alarm in the remote monitor center window. The rest of the paper is organized as follows. In Section 2, we describe the proposed system architecture. In Section 3, we describe different scenarios. In Section 4, we present the performance and discussion. Finally, the conclusions and future work are discussed in Section 5.

PROPOSED SYSTEM ARCHITECTURE

The WoT monitoring platform combines the WSN and the CAN bus technology to achieve comprehensive and timely monitoring and early warning of underground environmental parameters; in addition, the WoT technology is used to achieve remote monitoring of underground parameters in the remote monitor center. The entire system makes the abilities of the various underground sensors available to the application layer on the access gateway side and the operational system side via the Web technology; in addition, the system encapsulates the open interfaces via the REST-based framework and allows the application layer to access resources via HTTP. Figure 1 shows the overall architecture of the entire system.

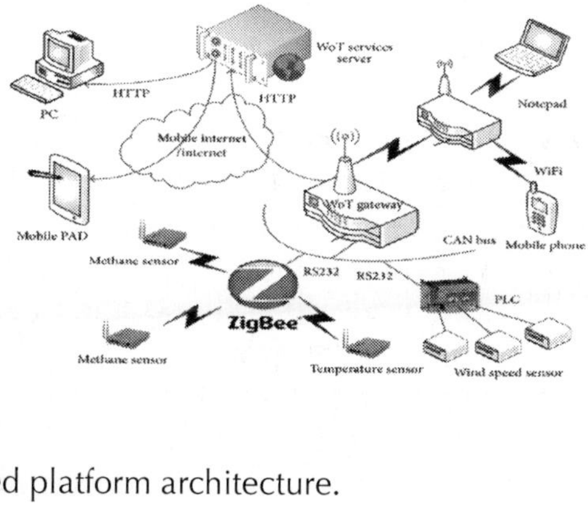

Figure 1: Proposed platform architecture.

The service gateway of the WoT is the core node, which can provide Web services, enabling applications to directly access the gateway in the Web form to communicate with the sensors/controllers.

The WoT service server provides WoT services to WoT applications and also provides such management functions as identification, routing, and registration for the WoT service gateway. Hence, applications based on this architecture can obtain different WoT services by accessing the WoT service gateway and the WoT service platform.

WoT applications comprehensively and timely monitor and provide intelligent early warnings on the underground environment, the production data, and the operating state of the equipment. In addition, all types of parameters are collected and transmitted to the remote monitor center for analysis to provide decision-making information for clients.

Underground Wireless Sensor Network Deployment

The ZigBee protocol supports star topology and peer-to-peer topology. In practice, wireless sensor nodes are not spread randomly, nor are they a self-organized, centerless network. In this proposal, the tree cluster structure was used to increase the stability and reliability of the network [16, 17]. The tree cluster structure is an application form of the peer-to-peer network topology as shown in Figure 2. In this proposal, the monitor center serves as the personal area network (PAN) coordinator, that is, the cluster head with a cluster labeling (CLD) of 0. All the fixed nodes serve as fixed cluster head nodes (CLH), that is, PAN coordinators. For the convenience of integrating with the location, every fixed node establishes a PAN with a fixed PAN identification (PAN ID) during network initialization and sends beacon frames to the adjacent device by broadcasting. The candidate device that received the beacon frames applies to the cluster head for joining the network. If the head cluster permits the application, then the cluster head will join its adjacent table as a child node. In addition, the device that asked to join will join the head cluster as the parent node to its adjacent table, such that the device becomes a slave device of the network. Once the capacity of the network reaches a certain limit, the head cluster will designate one slave device as the head cluster of another cluster network. Subsequently, nodes from an even larger range will join one by one, expanding the network successively until all the nodes covered by the network have joined to form a multicluster network (Figure 2).

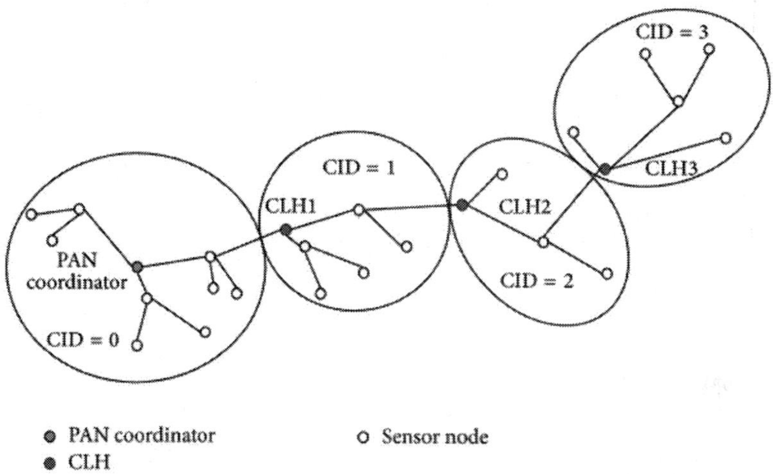

Figure 2: Tree cluster network.

Under normal situations, the node network process is as follows: when a ZigBee network is in initialization, a node is usually in a dormant state and wakes up at a random time point. After the node wakes up, the first step is to search for a network in its communication range. If a network is found, the node will select a parent device (a node that has already joined the network), based on the obtained network information, to apply to for permission to join the network and will await the request response from the parent node; if no network is found, then the node will declare itself the PAN coordinator and establish a network, accepting applications for permission to join the network as the parent device. After receiving an application to join the network, the parent device will make a decision on whether to approve the application based on the request information; if the parent device approves the application, then the parent device will send a request to inform the child device. After the child device receives the response message, it will obtain a network address assigned by the parent device as the only identifier within the network. The node has then successfully joined the network. After joining the network, the node will broadcast beacon frames as a coordinator and simultaneously accept requests from new nodes to join the network. Thus, through the level-by-level assignment of short addresses, all the nodes in the network-covered region will form a tree cluster network topology. Figure 3 shows the network setup process in the peer-to-peer wireless network nodes.

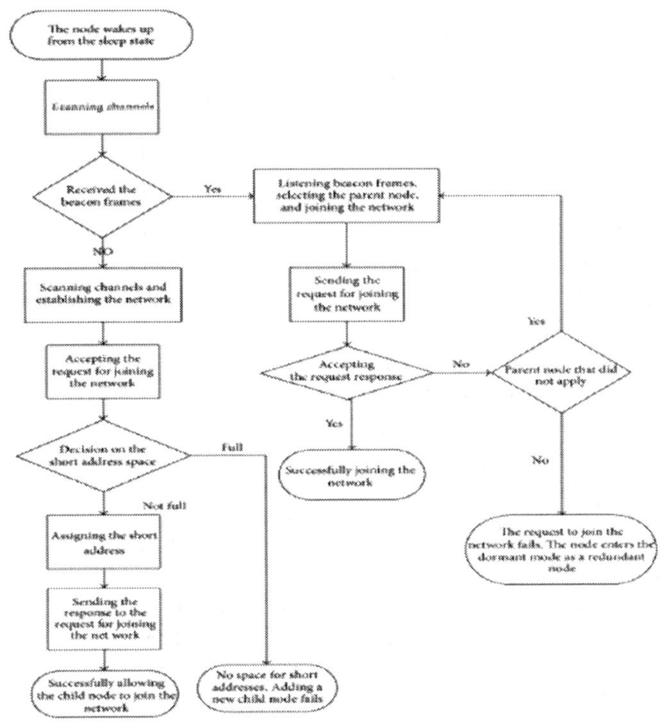

Figure 3: Network process setup process.

First, the coordinator scans the channels and detects energy, after which it establishes the network and accepts requests for permission to join the network from other nodes as the parent node. After the parent node receives a request to join the network, it first determines whether to allow the node to join the network; if the node is allowed to join the network, then the parent node will send a request response to inform the child node and assign a network address for the child node as its only identifier within the network. The node has then successfully joined the network. After joining, if the node is a router, it can in turn receive requests to join the network from new nodes. Through the level-by-level assignment of short addresses, all nodes in the monitor region will join the network. During the process of a node joining the network, a pair of nodes that communicate with one another form a parent-child relationship—the node that has already joined the network is the parent node, while the other node is the parent node's child node.

In coal mines, miners must constantly move due to work requirement. The ZigBee nodes carried on the miners move within, join, and exit the network. In addition, further mining results in the corresponding expansion of the network, which causes dynamic variation of the network topology. Nodes exchange data through the multihop data forwarding mechanism, which requires a special routing protocol for packet forwarding. In the proposed system, the data interaction primarily occurs between the monitor center and each monitoring node. The majority of data flow is from each monitoring node to the monitor center, followed by the instruction data sent from the monitor center to the devices; there is little data exchange between monitoring nodes. Therefore, the routing selection is optimized to some extent based on this feature. The data sent to the monitor center directly selects an upward vertical route along the cluster tree, while the data sent from the monitor center to a certain node directly selects a downward vertical route along the cluster tree; route discovery and route selection are only begun for the small amount of data exchange between nodes. When the network layer receives data from the upper layer, it first determines whether the transmission is a broadcast frame; if so, then it processes the data according to the broadcast frame procedure. The network layer determines whether the data should be sent to the monitor center; if so, the network layer uses its parent node as the route for the next hop; if not, then the network layer starts route selection or sends the data based on the known route for the next hop. If the node neither has the route discovery capability nor executes route selection, then it performs route selection along the tree. If the data are sent from the monitor center to a monitoring node, the data are first sent to the CAN bus. A fixed node checks whether the data belongs to its own PAN ID section; if so, then the node selects a downward vertical route along the cluster tree, and the data successively hop to the target node. If the data are an interaction between monitoring nodes, route discovery and route selection may be necessary.

WoT Gateway for Underground Sensors Networks

Underground sensor networks nodes must also address data delay tolerance, loss tolerance, distributed filtering, time synchronization, security, real time, low power, and other situations. The sink node then gathers the data and transmits them to a smart control gateway with better processing, storage, and communication capabilities than the sensor nodes. The smart control gateway receives data from sensor network nodes and instrument devices and accepts remote control commands from the external network. At the same time, the smart gateway control gateway has an embedded database to store data received from the sensor network nodes and instrument devices, which facilitates data management. For the higher level monitoring and control applications, any type of device looks the same, even if the underlying implementation differs. The framework for the WoT gateway is shown in Figure 4.

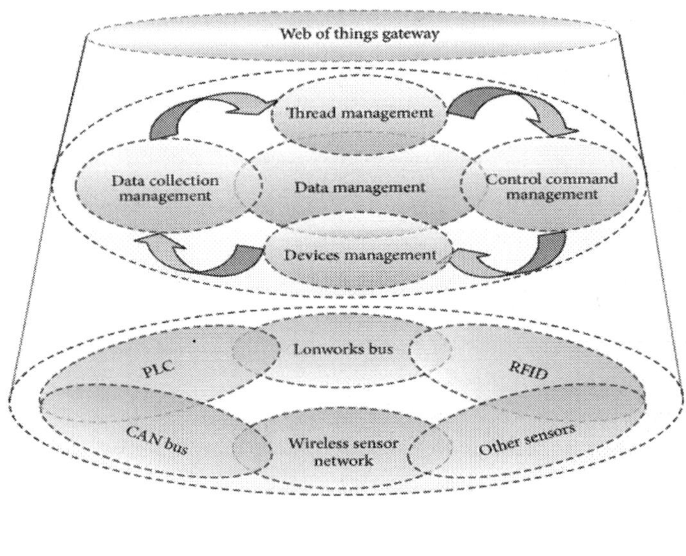

Figure 4: Web of Things gateway framework.

As seen in Figure 4, for underground physical devices that are not supported by HTTP over TCP/IP and that do not have an IP stack, the smart control gateway is responsible for translating the web request into a protocol understood by the physical device. The smart control gateway transforms the special communication protocol to general HTTP. It receives control messages and translates them to commands understood by the underground physical devices. Similarly, it collects data from the devices and forwards them to the monitoring center. Here, it contains the device management, data collection management, control command management, and protocol management.

The data collection management converts and parses data among different underground dedicated communication protocols. Sensors transmit data to a sink node, and the sink node receives the data and transmits it to the data collection management by the serial port. Further details of the control flow for data collection are illustrated in Figure 5.

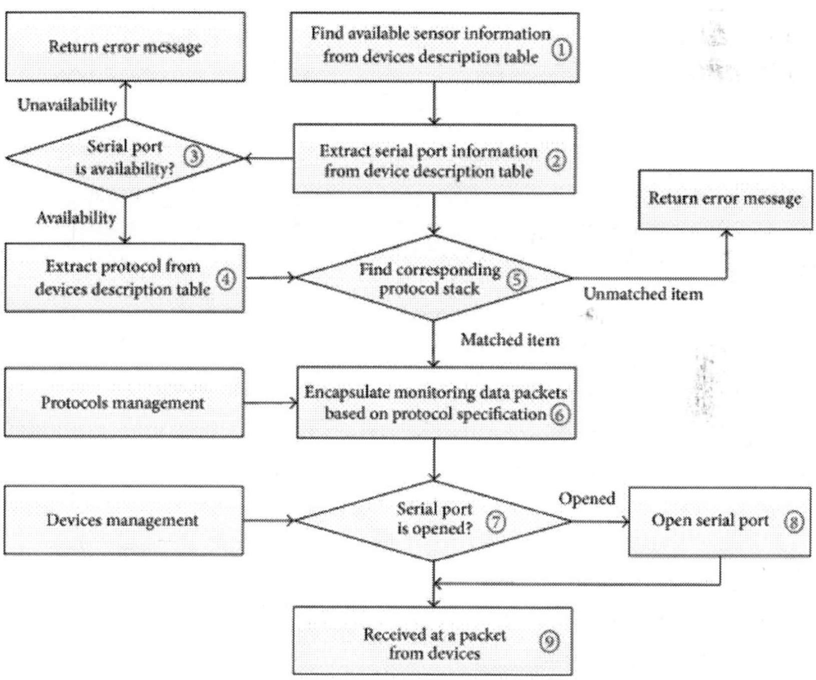

Figure 5: The control flow for data collection.

Step 1: Find available information for sensor devices from device description table. The STATUS field can equal 1, which means the device is available, or zero, which means the device is unavailable.

Step 2: Extract the serial port COM_ID from the matched sensor devices, which indicates which serial port ID should be used for the sensor devices.

Step 3: Judge whether the serial port COM_ID can be used, which can be determined based on the COM_ID from the serial port description table. If there are no matched items, or if there is a matched item but the STATUS field is equal to zero, which means this serial port is unavailable, then return an error message.

Step 4: Extract the protocol information for the devices and extract the DEVICE_PROTOCOL from the matched device items.

Step 5: Search the protocols from the protocol management module for the DEVICE_PROTOCOL protocol. If there is no matching protocol item, then return an error message.

Step 6: Encapsulate the data packet based on the protocol specification.

Step 7: Judge whether the serial port COM_ID is opened after the protocol data packet is encapsulated.

Step 8: Open the serial port, and if the serial port COM_ID is not opened, then open this serial port.

Step 9: Send the protocol data packets, and once the data channel is established when the serial port is opened, receive the data packet from the devices.

The control command management must perform any action commands. Automation control must send the command "ON" to turn on switches and the command "OFF" for shutdown. The control command management will analyze and parse the URL and send the commands to the devices. Further details of the control flow for command control are illustrated in Figure 6.

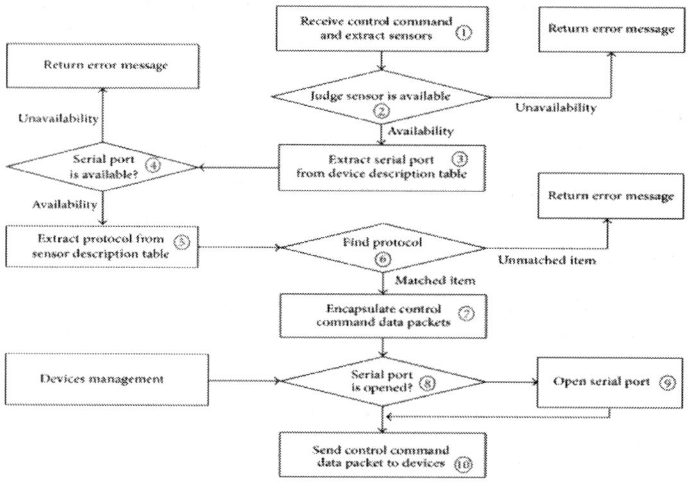

Figure 6: The flow for command control.

Step 1: Receive the control commands. The control command contains the name of the sensor device as NAME, PLC_ID, SENSOR_ID, and VALUE.

Step 2: Judge whether the device can be used based on the PLD_ID and SENSOR_ID, which are obtained from the device description table. If there are no matched items, or if there is a matched item but the STATUS field is equal to zero, which means this serial port is unavailable, then return an error messages.

Step 3: Extract the serial port COM_ID from the matched devices, which will be the serial port used for data access for the devices.

Step 4: Judge whether the serial port COM_ID can be used, which can be determined based on the COM_ID from the serial port description table. If there are no matched items, or if there is a matched item but the STATUS field is equal to zero, which means this serial port is unavailable, then return an error message.

Step 5: Extract the protocol information for the underground physical devices and extract the DEVICE_PROTOCOL from the matched device item.

Step 6: Search the protocols from the protocol management for the DEVICE_PROTOCOL protocol. If there is no matched protocol item, return an error message.

Step 7: Encapsulate the data packet based on the protocol specification.

Step 8: Judge whether the serial port COM_ID is opened after the protocol data packet is encapsulated.

Step 9: Open the serial port, and if the serial port COM_ID is not opened, then open this serial port.

Step 10: Send the protocol data packets, and once the data channel is established when the serial port is opened, send the control command data packet to the devices.

The device management is responsible for the interaction and management of devices, treating all the devices in the same manner, and it is responsible for sending and receiving messages to and from the devices. It deals mainly with low-level communication and is responsible for the interaction between the upper layers and the device threads. The device management maintains a high-level view of devices registered at the smart control gateway by using the device abstraction. It maintains a list of all the devices connected to the smart control gateway. The device management periodically sends a message to maintain the device list. If a new sensor device attempts to connect to the smart control gateway by sending a request message, when the request message is received successfully, it will immediately return the response message. Meanwhile, the newly added sensor device will be added to the device list for the smart control gateway and will also send a second response confirmation message to the sensor devices. Therefore, by sending a request to the device management with the device driver parameter class and a unique device identifier ID, new instances of the driver instances can be invoked, and the new sensor device drivers can be loaded through dynamic class loading. In particular, the device management receives a response message from the sensor device, which will generate a single URL address and the corresponding resource representation. However, if the response message is not received within five minutes because of the power consumption or abnormal work, the sensor device will be removed from the device list, and the corresponding URL will be destroyed.

The data management is responsible for real time management of data from the physical devices. It is necessary to collect the environmental data on the underground coal mine, such as methane gas, temperature, humidity, air pressure, wind speed, wind direction, and water level for each environmental parameter through the underground instrument and sensor network, then pass the data to the smart control gateway through the serial port and store those data in the database SQLite, which is a SQL database engine that is independent, embeddable, and zero-configuration, and occupies very few resources. The application can operate the tables in the SQLite database with different requirements.

The thread management is responsible for the device thread lifecycle management, which runs constantly, maintains all the threads for the smart control gateway, and initializes all the components of the smart control gateway. Once the smart control gateway discovers a new physical device, a new thread is automatically created, and inside every thread for physical devices, information is stored concerning the device description and a list of the services supported by the devices.

RESTful Remote Monitoring for Coal Mine

The underground coal mine sensing equipment is connected to the gateway via the communication network. The capabilities of the equipment are encapsulated by the proxy/gateway and are made accessible via the Web services on the gateway. The REST services use the four standard methods of HTTP to complete operations on resources. The operating targets are the resources marked by URLs. Every message is independent. This mode fully utilizes the characteristics of the existing HTTP. Figure 7 shows the request/response process.

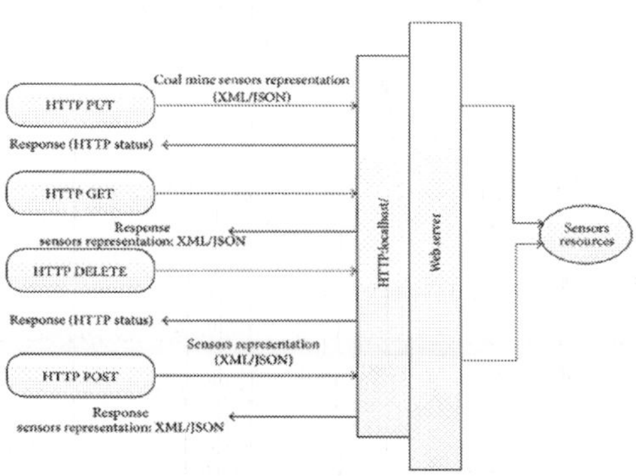

Figure 7: Request and response of REST services.

We classify transmitted messages into two types: data messages and downlink control messages. Due to the open interfaces encapsulated by REST style-based WoT services, HTTP is used to encapsulate both types of messages; in addition, both types of message support XML and JavaScript object notation (JSON) data exchange formats. Because HTTP serves as the application protocol, request messages will include both messages for operating the resources and the messages for transmitting resources. The resources are queried, modified, increased, and deleted via the HTTP methods GET, POST, PUT, and DELETE, respectively.

When an application sends a query request for data messages to the WoT architecture, the request mode-GET-of the HTTP head reflects the query to the WoT resources. Then, a returned data message includes a message head and a message body. Table 1 lists the field definitions of the response data structure.

Table 1: Field definitions of data structure

Fields	Field Type	Field definition
Data	XML/JSON	Data information
Datapoints	XML/JSON	Data list, –1 or multiple data values
Value	String	Data
Datatype	String	One of data attributes, indicating data type
Unit	String	One of data attributes, indicating data unit
Time	Time	One of data attributes, indicating time for obtaining data

Box 1 is an example of an XML data format-based data query message body. In this example, the response was the data on methane concentration at 16:22:10 on 2013-05-10.

```
<Data xmlns:xsi="http://www.w3.org/2001/XMLSchema-instance"

Xmlns:xsd=http://www.w3.org/2001/XMLSchema svc_ID="222">

<Datapoints>

<value datatype="methane" Unit="Con"

Time="2013-05-10T16:22:10">5</value>

</Datapoints>

</Data
```


Box 1

For the command control messages, when an application sends a command control request to the WoT gateway, the request mode of the HTTP head, POST, reflects the modification of the WoT resources. Modifying a resource will require new data information. Therefore, the request packet also includes the information on the new resource, which is attached to the end of the request head. Table 2 lists the field definitions of the control information.

Table 2: Field definitions of control information

Fields	Field Types	Field definitions
Device	XML/JSON	Data information
Operation	XML/JSON	Data list
Parameters	XML/JSON	Parameter list, 1 or multiple
Parameter	String	New parameter value
Name	String	One of the parameter attributes, indicating the name of the parameter

Box 2 is an example of an XML data format-based downlink control message. In this example, the device that needed to be modified was the device with a Svc ID 111, and the first channel of the device was opened.

```
<Device Svc_ID="111" version="1.0">
<Operation_ID="ON">
<parameters>
<parameter name="channel">02</parameter>
</parameters>
</Operation>
</Device>
```


Box 2

For operation requests, a response message only includes the message head; whether the request was successful is indicated by the message object code.

RESTful Web services use the 4 standard methods of HTTP to complete the operations on resources. The operating targets are the resources identified by URLs. The detailed realization of the REST design of each operation is discussed in detail as follows. The design follows the standard of Java API for RESTful Web Services (JSR-311) (JAX-RS). The standard defines a number of annotations, which are used to encapsulate a class of resource or a class of Java into a Web resource. The main annotations used here include the following: @ Path (annotating the relative path of a resource class or method); @ GET, @PUT, @POST, and @DELETE (annotating the HTTP request type of the method); and @Produces and @Consumes (annotating the multipurpose internet mail extensions (MIME) media type that is returned to the client and introduced by the server). The following are the parameter types: @PathParam, @QueryParam, @HeaderParam, @ CookieParam, @MatrixParam, and @FormParam, which annotate the parameters of the method from different locations of an HTTP request. For instance, @PathParam is from the path of the URI; @QueryParam is from the query parameter of the URL; @HeaderParam is from the head information of an HTTP request; and @CookieParam is from the cookie of an HTTP request.

The resource register interface mainly registers the basic resource information of each underground sensor to the WoT gateway server, enabling the client to obtain all the necessary information for operating services on the WoT gateway server. Box 3 gives the detailed definition of the interface.

```
@POST
@Path ("/{servicename}")
@Consumes ("application/x-www-form-urlencoded")
String addService(@PathParam("servicename") String servicename,
@FormParam("user") String user, @FormParam("type") String type,
@FormParam("descrip") String descrip, InputStream servicestream);
```

Box 3

Here, @POST indicates that the service register interface submits a request in the form of HTTP POST. @Path("/{servicename}") defines the URI of the operation; the base address of the interface operation is http://localhost:8080/; the path can be directly defined with a variable; and the final URI is base address + path.@Consumes("application/x-www-form-urlencoded") indicates the request type, that is, the transmission type of the parameter of the request that the server received.

The resource update interface mainly updates the state of the resource of each underground sensor. Due to the effects of variation in the original requirements, the data source address, and the parameter configuration. Box4 shows the definition for the interface.

```
@PUT
@Path ("/{servicename}")
@Consumes ("application/x-www-form-urlencoded")
```

String updateService(@PathParam("servicename") String servicename,
InputStream servicestream);

Box 4

Here, @PUT indicates that the service update interface submits requests in the form of PUT. @Path("/{servicename}") is used to configure the Web address that the interface accesses; the base address is http://localhost:8080/mashup, and the final accessed Web address is the base address + the configured path, http://localhost:8080 /servicename. @ Consumes(application /x-www-form- urlencoded) is used to configure the format of the parameters transmitted to the server, which is mainly used to configure the service stream.

The resource deletion interface mainly deletes the resources of the sensors that cannot be called to clear useless information from the database on the WoT gateway server. The definition of the interface is as shown in Box5.

@DELETE
@Path ("/{servicename}")
Boolean deleteService(@PathParam("servicename") String servicename;

Box 5

Here, @DELETE indicates that the resource deletes interface requests called in the form of HTTP DELETE. @Path("/{servicename}") is used to configure the Web address that the interface accesses; the base address is http://localhost:8080, and the accessed Web address is the base address + the configured path, http://localhost:8080/ servicename.

The main functions of the resource query interface are to search for the services that meet the requirements among the services that have been registered in the server according to the client's specified query parameters, such as the registered service client, service type, and service name, and to return the service list to the client. The definition of the interface is as shown in Box 6.

```
@GET
@Path ("/services")
List<Service> queryService(@QueryParam("user") String user;
@QueryParam ("type") String type, @QueryParam("servicename")
String servicename);
```

Box 6

Here, @GET indicates that the service query interface submits requests in the form of HTTP GET. @Path("/services") defines the Web address accessed by the query interface. Because the query interface will return multiple service lists according to the query conditions, it is not possible to use the base addresses used in the four aforementioned interfaces + service name as the web address. A new path, /services, is added to the path variable. The final accessed web address of the service query interface is http://localhost:8080/services.

WOT MONITORING FOR COAL MINE SCENARIOS

In the first example, an all-around monitoring viewgraph of the underground environment is provided in Figure 8, which mainly includes displays of real-time monitor data. Therefore, clients can browse the environmental parameters, such as coal mine gas, carbon monoxide, negative pressure, temperature, and wind speed, as well as the operating state of such equipment as the blowers, the ventilators, and the air doors at any time at any location. Overlimit warning messages are displayed in bright red. Reports of various monitor data can be generated for statistical analysis and printing. Reports include the analog data statistical report, the analog warning statistical report, the analog power outage report, the analog feed anomaly report, the variation of the state of the switch signal report, the switch signal warning report, the switch signal power outage report, the switch signal feed anomaly report, and the monitoring equipment failure report.

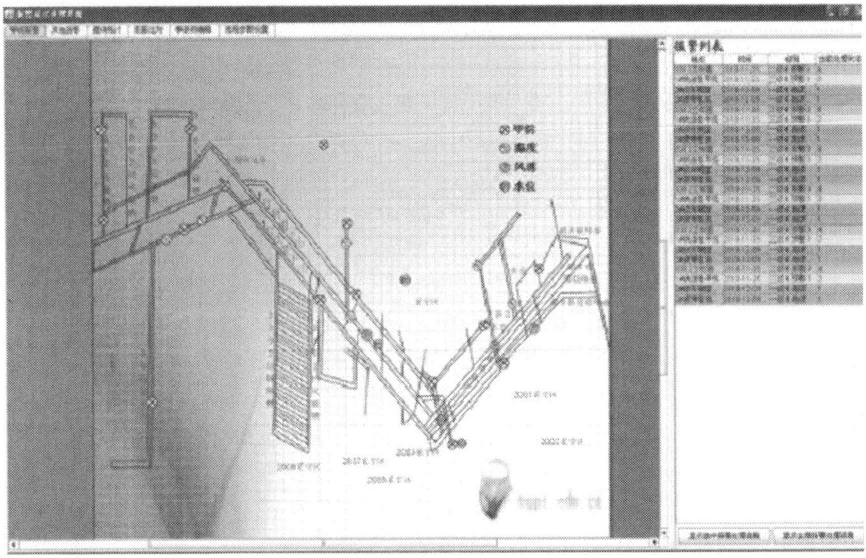

Figure 8: All-around monitoring viewgraph for coal mine.

In the second example, coal mine safety information is vibrant, dynamic, and spatial location-related information. The geographic information system (GIS) collects, stores, updates, analyzes, transmits, and inquires about geographical information; the GIS is a widely used tool in the management process. The GIS can be used to provide the coal mine supervisors and the higher authority department with fast, accurate, visual, and systematic information with pictures and text for coal mines. Thus, GIS technology plays a more and more important role in remote monitoring systems for coal mine safety. Figure 9 shows the graphical integrated view for a monitoring and control system for a coal mine based on mobile PDA devices.

Figure 9: Mobile GIS for coal mine.

After the mobile terminal is connected to the remote server of the coal mine, the communication and data exchanges between the two are completed. The coal mine on-site measuring point data on the mobile terminal is established via the mobile network. The diverse management applications of the GIS space monitoring services, such as correlation, positioning, and real-time alarm, alter the traditional situation, in which coal mine safety production management systems are all PC-based and provide a brand new method for coal mine safety production monitoring and management.

In the third example, suitable cameras are installed at each substation according to the actual situations (e.g., lighting, appearance of the pit and the mine, locomotive equipment). Cameras are installed on the cradle head and controlled by the ground monitor center (rotating, zooming, and adjusting the angle) to ensure all-around monitoring on the underground observation points. Figure 10 shows the mobile video monitoring scenario for the underground coal mine.

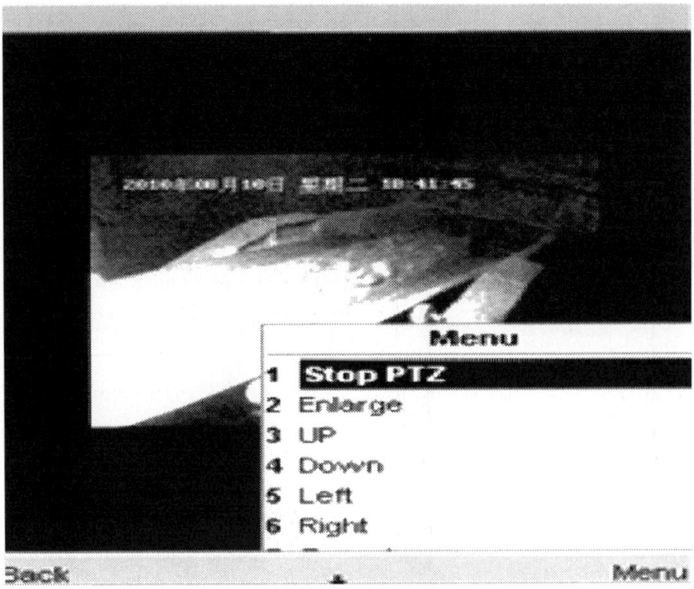

Figure 10: Mobile video monitoring for coal mine.

The mobile phone video monitoring system can transmit the videos of the monitored objects collected by the cameras to smartphones in real time via the 3G data network to allow the supervisors to check the state of the monitored objects at any time at any location. When viewing the monitoring videos, supervisors can control the cradle head of the cameras via their smartphones (e.g., zooming, focusing, and adjusting the direction).

PERFORMANCE AND DISCUSSION

In this section, we attempt to evaluate our system experimentally. We created different scenarios to test our system under real-time and heavy workload circumstances. Through these tests, we measure the performance and reliability of our system. The experimental setup used to evaluate our framework is as follows. The remote monitoring software was installed on a PC server (Intel Xeon E5420 2.5 GHz with 10 G RAM). Around the REST smart gateway, we deployed 30 Mica2 motes. These sensor nodes were distributed on a tunnel wall approximately 10 meters wide and 5 meters high. Additional sensor devices were also deployed.

Experiment 1: The service registry interfaces and the service query interfaces of simple object access protocol (SOAP) and REST were tested, respectively. The service registry already included the database operation and the service script file operation, while the service query operation only included the database query operation. The access modes were as follows: RPCServiceClient and HttpClient client terminals were used for access five times, and the service access times were recorded. The service access time refers to the duration from the time when the client terminal requests the service to the time when the server response is received. Figure 11compares the results.

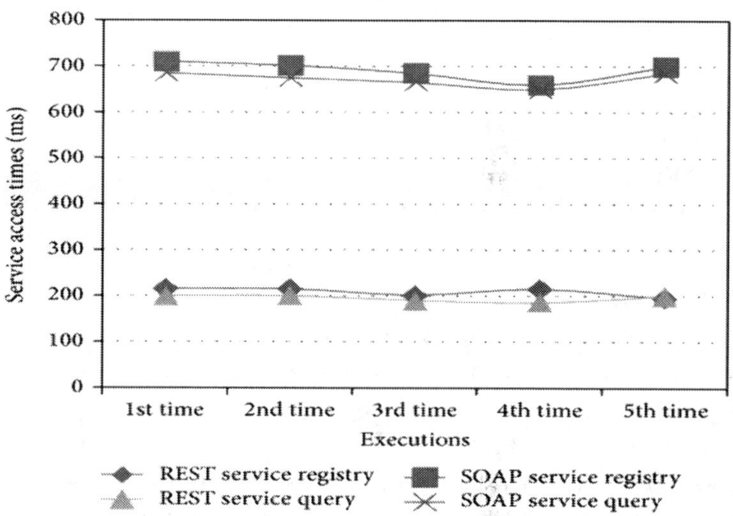

Figure 11: Comparison for response time between REST and SOAP.

It can be seen in Figure 11 that the access efficiency of the REST Web services was significantly higher than the access efficiency of the SOAP Web services, mainly due to the following: it was necessary to construct SOAP messages when accessing the SOAP Web services; SOAP messages needed to include the web addresses of the services to be accessed and the corresponding operation and parameter information; in addition, the number of sent request messages was far greater than the number of REST Web service-based HTTP requests for access. HTTP received a significantly optimized network transmission protocol; thus, the response efficiency of HTTP was higher than that of SOAP messages.

Experiment 2: Based on the performance test results, the video monitoring service was taken as an example, and a table of comparison test results was provided, mainly for comparison of the average completion time parameters under different concurrent numbers; the result is shown as in Figure 12.

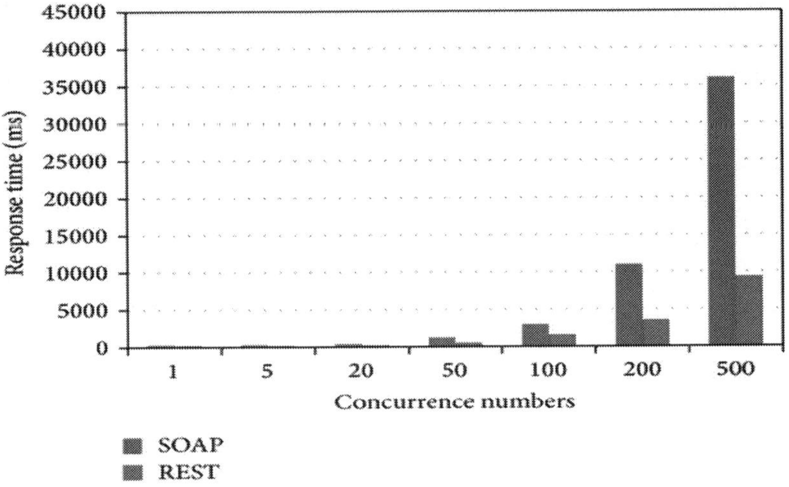

Figure 12: Comparison response time bet SOAP and REST.

Figure 12 shows that with the increasing concurrent number, the time parameters of the two types of services exhibited an increasing nonlinear trend. When the concurrent numbers were the same, all the time parameters of the SOAP Web services were greater than the time parameters of the REST Web services. With the increasing concurrent number, the difference between the time parameters of the two types of services became more and more significant. Therefore, the REST Web services were superior to the SOAP Web services in terms of performance and efficiency, which was mainly due to such advantages as the simple packet structure of the REST Web services and the HTTP cache that gave the REST Web services higher efficiency under the concurrent conditions.

Experiment 3: When implementing the REST Web services, either JSON or XML can be selected as the data format. The Windows communication foundation (WCF) framework can serialize and deserialize these two types of data formats. We took video monitoring as an example and conducted performance tests on the REST Web services established using the JSON and XML data transmission formats, respectively. Figure 13compares the results.

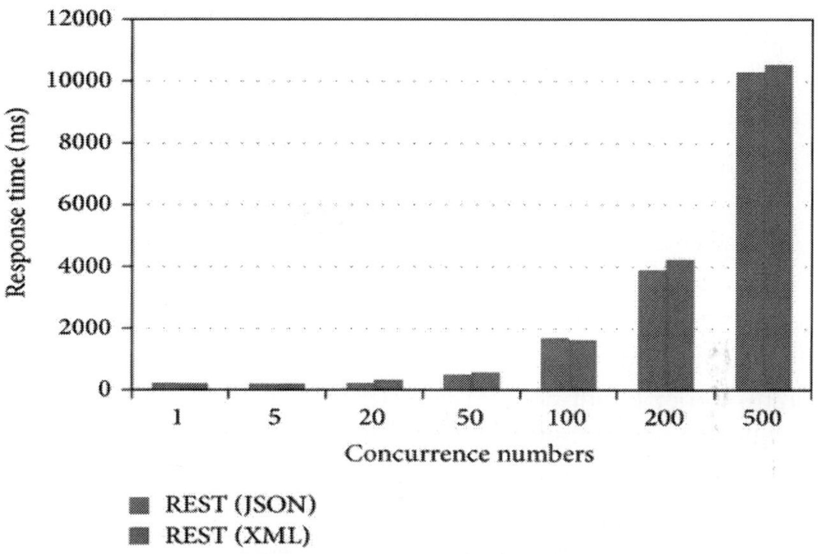

Figure 13: Comparison response time between JSON and XML REST Web services.

As seen in Figure 13, when the concurrent numbers were the same, the minimum time of the JSON data format was slightly shorter than the minimum time of the XML data format; however, the maximum completion time of the JSON data format was generally longer than the maximum completion of the XML data format; in addition, there was no significant difference between the two types of services in terms of the mean time. With the increasing concurrent number, there was only a slight difference between the two types of services in terms of the time parameter. Overall, the JSON data format was slightly better than the XML data format in terms of performance.

CONCLUSIONS AND FUTURE WORK

This paper proposes a WoT-based remote monitoring system that takes full advantage of wireless sensor networks in combination with the

CAN bus communication technique abstracts the underground sensor data and capabilities into WoT resources to offer services using REST style. The different remote monitoring scenarios for coal mine safety are illustrated, and the system performance is also tested and analyzed. In the future work, we plan to improve and optimize the deployment for the wireless sensor network underground coal mine and also to improve the coal miner localization precision using ZigBee wireless sensor network.

ACKNOWLEDGMENTS

This research is supported by the National Grand Fundamental Research 973 Program of China under Grant nos. 2011CB302506 and 2012CB315802, National High-tech R&D Program of China (863 Program) under Grant no. 2013AA102301, National Natural Science Foundation of China under Grant no. 61132001, and Program for New Century Excellent Talents in University (Grant no.NCET-11-0592).

REFERENCES

1. M. Li andY. Liu, "Underground coal mine monitoring with wireless sensor networks," ACM Transactions on Sensor Networks, vol. 5, no. 2, pp. 1–29, 2009.
2. P. Misra, S. Kanhere, D. Ostry, and S. Jha, "Safety assurance and rescue communication systems in high-stress environments: a mining case study," IEEE Communications Magazine, vol. 48, no. 4, pp. 66–73, 2010.
3. H. Jiang, J. Qian, and W. Peng, "Energy efficient sensor placement for tunnel wireless sensor network in underground mine," in Proceedings of the 2nd Conference on Power Electronics and Intelligent Transportation System (PEITS '09), pp. 219–222, December 2009.
4. E.-J. Ding, C.-N. Wang, and Q. Zhou, "Wireless model and deployment of sensor networks in the mine," in Proceedings of the International Conference on Information Acquisition (ICIA '07), pp. 538–542, Seogwipo, Republic of Korea, July 2007.

5. X. Wang, X. Zhao, Z. Liang, and M. Tan, "Deploying a wireless sensor network on the coal mines," inProceedings of the IEEE International Conference on Networking, Sensing and Control (ICNSC '07), pp. 324–328, London, UK, April 2007.
6. V. Guangxiang and G. Shengyi, "Application and practice of digital coal mine safety monitoring system," China Coal, vol. 30, pp. 52–54, 2004.
7. Web Services Discovery and Web Services Devices Profile (WS-DD), OASIS Open, 2009,https://www.oasis-open.org/committees/ws-dd/.
8. P. Spiess, S. Karnouskos, D. Guinard et al., "Soa-based integration of the internet of things in enterprise services," in Proceedings of the IEEE International Conference on Web Services (ICWS '09), pp. 968–975, July 2009.
9. D. Savio and S. Karnouskos, "Web-service enabled wireless sensors in SOA environments," inProceeding of the 13th IEEE International Conference on Emerging Technologies and Factory Automation (ETFA '08), pp. 952–958, Hamburg, Germany, September 2008.
10. C. Pautasso, O. Zimmermann, and F. Leymann, "RESTful web services vs. big web services: making the right architectural decision," in Proceedings of the 17th International Conference on World Wide Web (WWW '08), pp. 805–814, New York, NY, USA, April 2008.
11. F. Rosenberg, F. Curbera, M. J. Duftler, and R. Khalaf, "Composing RESTful services and collaborative workflows: a lightweight approach," IEEE Internet Computing, vol. 12, no. 5, pp. 24–31, 2008.
12. D. Guinard, "Towards the web of things: web mashups for embedded devices," in Proceedings of the International World Wide Web Conference (WWW '09), Madrid, Spain, 2009.
13. J. I. Fernández-Villamor, C. Á. Iglesias, and M. Garijo, "Microservices lightweight service descriptions for REST architectural style," in Proceedings of the 2nd International Conference on Agents and Artificial Intelligence (ICAART ‹10), pp. 186–189, January 2010.

14. D. Guinard, V. Trifa, and E. Wilde, "A resource oriented architecture for the web of things," inProceedings of the 2nd International Internet of Things Conference (IoT ‹10), December 2010.
15. S. Makelainen and T. Alakoski, "Fixed-mobile hybrid mashups: applying the REST principles to mobile-specific resources," in Proceedings of the International Workshops on Web Information Systems Engineering, pp. 172–182, 2008.
16. P. Jurcik, R. Severino, A. Koubaa, M. Alves, and E. Tovar, "Real-time communications over cluster-tree sensor networks with mobile sink behavior," in Proceedings of the 14th IEEE International Conference on Embedded and Real-Time Computing Systems and Applications, pp. 401–412, Kaohsiung, Taiwan, August 2008.
17. P. Jurčík and Z. Hanzálek, "Simulation study of energy efficient scheduling for IEEE 802.15.4/ZigBee cluster-tree wireless sensor networks with time-bounded data flows," Tech. Rep. TR-0911-1, Department of Control Engineering, CTU, Prague, Czech Republic, 2009, http://rtime.felk.cvut.cz/tdcs.

Chapter 6

Remotely Sensed Image Retrieval Based on Region-Level Semantic Mining

Tingting Liu[1], Liangpei Zhang[2], Pingxiang Li[2], and Hui Lin[3]

[1]Chinese Antarctic Center of Surveying and Mapping, Wuhan University, Wuhan, China

[2]The State Key Laboratory of Information Engineering in Surveying, Mapping, and Remote Sensing, Wuhan University, Wuhan, China

[3]Institute of Space and Earth Information Science, The Chinese University of Hong Kong, Hong Kong, China

ABSTRACT

As satellite images are widely used in a large number of applications in recent years, content-based image retrieval technique has become important tools for image exploration and information mining; however, their performances are limited by the semantic gap between low-level features and high-level concepts. To narrow this semantic gap, a region-level semantic mining approach is proposed in this article. Because it is easier for users to understand image content by region, images are segmented into several parts using an improved segmentation algorithm, each with homogeneous spectral and textural characteristics, and then a uniform region-based representation for each image is built. Once the probabilistic relationship among image, region, and hidden semantic is constructed, the Expectation Maximization method can be applied to mine the hidden semantic. We implement this approach on a dataset consisting of thousands of satellite images and obtain a high retrieval precision, as demonstrated through experiments.

INTRODUCTION

The information in remotely sensed images plays an important role in environmental monitoring, disaster forecasting, geological survey, and other applications. With the steadily expanding demand for remotely sensed images, many satellites have been launched, and thousands of images are acquired every day [1]. This leads to an exponential increase in the quantity of remotely sensed images in database. Therefore, how to retrieve useful images quickly and accurately from a huge and unstructured image database becomes a challenge.

Traditional image query techniques retrieve images by matching keywords [2], such as geographic location, sensor type, and time of acquisition. But, the content of the image which is much more important than attributes is not considered in these techniques [3]. In order to overcome this shortcoming of these techniques, image retrieval techniques are strongly focused on content-based image retrieval (CBIR). In a CBIR system, low-level features are used to represent image content and retrieve image from database, such as spectrum, texture, and shape [4-6]. Although low-level features can accurately be extracted by various methods, they cannot easily be utilized to describe user's perception of an image [7,8]. Semantic feature is a high-level hidden concept, which is meaningful to user's perception. The difference between low-level feature and high-level semantic feature, caused by the absence of a direct relationship between low-level features and high-level concepts [9-13], is called the "semantic gap" [7]. To narrow this gap, a semantic-based image retrieval system should be built, in which high-level semantic features can automatically be extracted from low-level image features.

Semantic feature mining is essential to semantic-based image retrieval technique. The process of semantic feature mining can be divided into two steps: low-level feature extraction and high-level semantic feature extraction. At present, most studies on low-level feature extraction are based on pixel characteristics. Li and Narayanan [14] identified ground-cover information based on spectral characteristics using supervised classification and extracted textural features by characterizing spatial information using Gabor wavelet coefficients. Li et al. [15] developed an approach based on pixel-level textural information to extract global semantic features. But, pixel does not facilitate understanding of the image, thus it is often replaced by region. It can be assumed that region-level description of visual information is more comprehensible for users than pixel-level image description. However, most of the existing high-level semantic feature mining methods are based on pixel-level features. Datcu et al. [16] and Daschiel and Datcu [17] developed a Bayesian classifier to retrieve images from a remotely sensed image database by approximating the probabilities of images belonging to different classes using pixel-level probabilities. Aksoy et al. [1] proposed a pixel-based Bayesian framework for a visual grammar to narrow the gap between low-level features and high-level concepts. Therefore, in this article, a novel approach is proposed to achieve region-level semantic feature mining. First, a region-level image content description is developed to facilitate users' understanding of image. Based on region-level features, a probabilistic relationship among image, region, and hidden semantic is developed. Then, the Expectation Maximization (EM) method is used to mine the hidden semantic features. Finally, remotely sensed image retrieval is performed using region-level semantic features.

The rest of the article is organized as follows. In Section 2, details of the region-level image representation are provided. In Section 3, semantic mining using the EM method is discussed. In Section 4, experiments are presented to demonstrate the effectiveness of region-level semantic features. Finally, conclusions are presented in Section 5.

REGION-LEVEL IMAGE REPRESENTATION

Region-level image representation includes the following components: image segmentation, regional information description, and codebook extraction. Figure 1 shows a flowchart of the region-level image representation process.

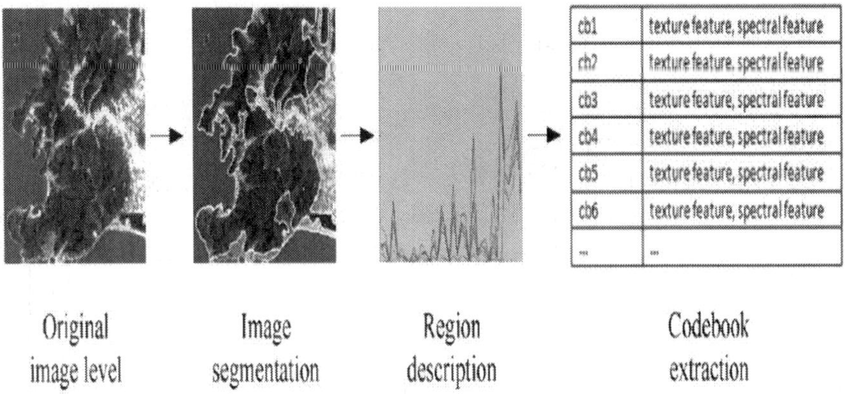

Figure 1: Flowchart of region-level image representation.

Image Segmentation

The JSEG algorithm [18] is a region-based segmentation method that provides robust segmentation results for a large variety of images and videos [19-21]. In this article, the JSEG algorithm is improved to make it applicable to multi-spectral remotely sensed image segmentation.

The JSEG algorithm consists of two parts: color quantization and spatial segmentation. Figure 2shows a schematic diagram of the original JSEG algorithm.

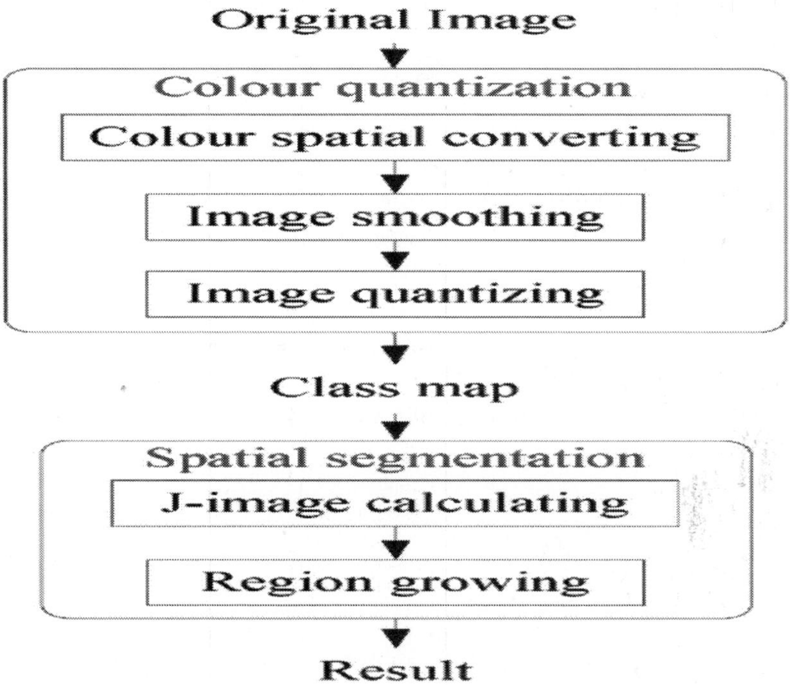

Figure 2: Schematic diagram of the original JSEG algorithm.

In color quantization step, the general Lloyd algorithm (GLA) [22] is used to quantize the image. In this algorithm, the distortion D can be represented as Equation (1)

$$D = \sum_i D_i \sum_i \sum_n v(n) \left\| x(n) - c_i \right\|^2, x(n) \in C_i \tag{1}$$

where C_i is the ith cluster in the image, c_i is the center pixel of cluster C_i, $x(n)$ and $v(n)$ are the color vector and the perceptual weight for pixel n, and D_i is the total distortion for cluster C_i.

Since multi-spectral Thematic Mapper (TM) images are used as experimental data, $x(n)$ is defined as $x(n) = \{a_{n1}, a_{n2}, \ldots, a_{nj}\}$ in this algorithm, where j is the number of bands in the image and a_{nj} is the value of the nth pixel in the jth band of the image.

In spatial segmentation step, region growing method is used to segment image based on J-image, in which a threshold controls region growing result. In this research, 0.4 is chosen as an empirical value.

Remotely sensed images present complex spatial arrangement and spectral heterogeneity. It has been demonstrated that combing spatial and spectral information can improve land cover information extraction from satellite image data [23]. Therefore, in this research, Normalized Difference Vegetation Index (NDVI) [24], Normalized Difference Built-up Index (NDBI) [25], and textural features are substituted for the original spectral features to increase land cover separability. NDVI provides a standardized method of assessing whether the land cover being observed contains live green vegetation or not; it can be calculated as Equation (2)

$$NDVI = \frac{(NIR - R)}{(NIR + R)} \tag{2}$$

where R and NIR stand for the spectral reflectance measurements acquired in the visible (red) and near-infrared regions, respectively. NDBI serves to compare urban areas with built-up areas between satellite images; it can be calculated as Equation (3)

$$NDBI = \frac{(MIR - NIR)}{(MIR + NIR)} \tag{3}$$

where MIR and NIR stand for the spectral reflectance measurements acquired in the middle-infrared and near-infrared regions, respectively. Texture reflects the local variability of grey level in the spatial domain and reveals the information about the object structures in the natural environment [26]. In this research, the mean texture is used and extracted using Grey-Level Co-occurrence Matrix (GLCM), which is commonly applied in statistical procedure for interpreting texture. Finally, the pixels in original image can be represented as Equation (4)

$$f = \{f_{NDVI}, f_{NDBI}, f_{texture}\} \tag{4}$$

As described above, a flowchart for the improved segmentation algorithm is shown in Figure 3.

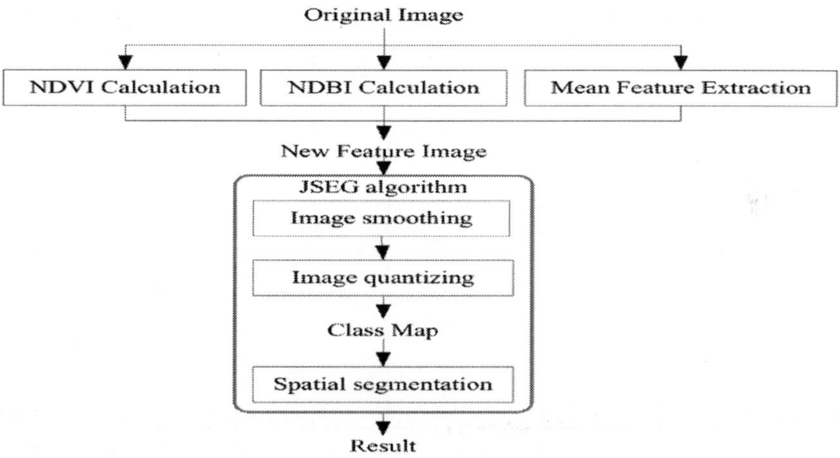

Figure 3: Flowchart of improved image segmentation algorithm.

Regional Information Description

In this research, regional information is described using spectral and textural features. Spectral feature is the original pixel value, and textural feature is extracted using GLCM. These two features are extracted separately for each region in all images.

GLCM is a commonly used method in texture analysis. It describes the frequency at which one grey tone appears in a specified spatial linear relationship with another grey tone in the area under investigation. Fourteen statistical parameters [27] can be extracted using GLCM. However, in the retrieval system, the more features, the lower the efficiency is. The correlation matrix of the eight common parameters, namely mean, variance, homogeneity, contrast, dissimilarity, entropy, second angular moment, and correlation, is presented in Table 1.

Table 1: The correlation matrix of the eight common parameters

	Mean	Variance	Homogeneity	Contrast	Dissimilarity	Entropy	ASM	Correlation
Mean	1.00	0.62	-0.51	0.58	0.77	0.55	-0.29	-0.13
Variance	0.62	1.00	-0.44	0.80	0.76	0.32	-0.25	0.04
Homogeneity	-0.51	-0.44	1.00	-0.46	-0.70	-0.39	0.74	0.14
Contrast	0.58	0.80	-0.46	1.00	0.87	0.30	-0.24	0.03
Dissimilarity	0.77	0.76	-0.70	0.87	1.00	0.56	-0.41	-0.04
Entropy	0.55	0.32	-0.38	0.31	0.56	1.00	-0.57	-0.34
ASM	-0.29	-0.25	0.74	-0.24	-0.42	-0.57	1.00	0.28
Correlation	-0.13	0.04	0.14	0.03	-0.04	-0.34	0.28	1.00

Liu et al. EURASIP Journal on Image and Video Processing 2012 2012:4, doi:10.1186/1687-5281-2012-4

The correlation matrix shows the correlation of the two parameters. The higher value indicates the higher correlation of the two parameters. In Table 1 there are two correlations over 0.8. The correlation value between contrast and variance is 0.80; the correlation value between dissimilarity and contrast is 0.87. These indicate that variance and dissimilarity are highly correlated with contrast. Therefore, variance and dissimilarity can be replaced by contrast, while the other six parameters describe textural features.

Codebook Extraction

After the images have been segmented into several parts, a number of regions are generated and stored in the database. It will be time-consuming to calculate the similarity between two regional features for all pairs of regions.

However, many regions on different images are very similar in terms of spectral and textural features. Therefore, GLA is used to classify the low-level features into a set of codes based on which a codebook will be generated (as shown in Figure 4).

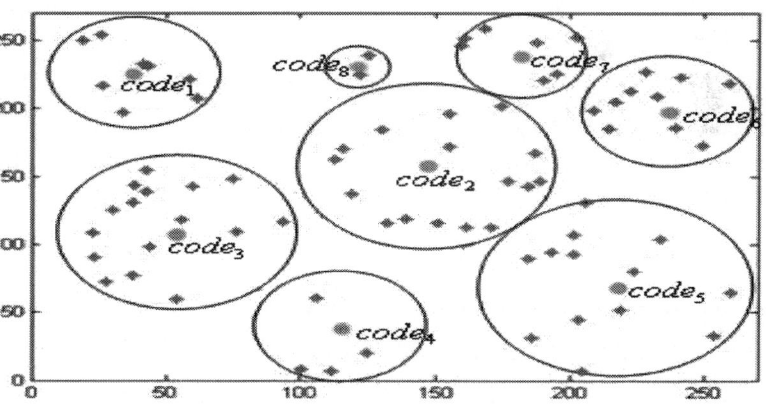

Figure 4: Schematic diagram of codebook extraction.

Figure 4 presents the principle of codebook extraction when image feature is two-dimensional feature space. In Figure 4, the blue point is a low-level feature, the black circle is a cluster, and the red point is the center of the cluster called code. Code_j is the mean of all features in corresponding cluster. All codes form a codebook. Then, each region can be represented by a code. For an image *I*, its *i*th region R_i can be represented by Code_j.

SEMANTIC FEATURE EXTRACTION

In this step, a probabilistic method is used to mine the relationship among semantic features, regions, and images automatically. Then the EM method [28,29] is used to analyze the relationship and extract the latent semantic concepts.

First, various parameters are defined as follows:

- *Image data*: d_j is an image in the database, $d_j \boxtimes \{d_1,\ldots,d_M\}$; *M* is the total number of images.
- *Regional feature data*: r_i is the *i*th region feature in the feature codebook, $r_i \boxtimes R = \{r_1,\ldots,r_N\}$, where *N* is the total number of regional features.
- *Hidden semantic features*: s_k is the hidden semantic feature, $s_k \boxtimes S = \{s_1,\ldots,s_K\}$, where *K* is the total number of semantic features.

where *j* is the number of images, $j \boxtimes \{1,\ldots,M\}$; *i* is the number of region features, $i \boxtimes \{1,\ldots,N\}$; *k*is the number of semantic features, $k \boxtimes \{1,\ldots,K\}$.

$P(d_j)$ denotes the probability that an image will occur in a particular image database. $P(r_i|s_k)$ denotes the class-conditional probability of region r_i given the hidden semantic feature s_k. $P(s_k|d_j)$ denotes the class-conditional probability of the hidden semantic feature s_k given a particular imaged_j. d_j and r_i are independently defined on the state of the associated hidden semantic feature. According to conditional probability formula, the joint probability of d_j and r_i can be described by Equation

$$P(r_i, d_j) = P(d_j)P(r_i|d_j) \tag{5}$$

Then, applying total probability formula, Equation (5) can be transformed to Equation (6):

$$P(d_j)P(r_i|d_j) = P(d_j)\sum_{k=1}^{K} P(r_i|s_k)P(s_k|d_j) \tag{6}$$

The class-conditional probability of semantic feature s_k, $P(s_k|r_i,d_j)$, depends on image d_j and region feature r_i. Using the Bayesian formula, this class-conditional probability can be described by Equation (7)

$$P(s_k|r_i,d_j) = \frac{P(r_i,d_j|s_k)P(s_k)}{P(r_i,d_j)} \tag{7}$$

Since d_j and r_i are independent, referring to Equation (6), Equation (7) can be transformed to

$$P(s_k|r_i,d_j) = \frac{P(d_j|s_k)P(r_i|s_k)P(s_k)}{P(d_j)\sum_{k=1}^{K} P(r_i|s_k)P(s_k|d_j)} \tag{8}$$

where l is the number of semantic features, l ⊠ {1,...,K}.

Referring to Bayesian formula, Equation (8) can be transformed to

$$P(s_k|r_i,d_j) = \frac{P(d_j|s_k)P(r_i|s_k)P(s_k)}{P(d_j)\sum_{k=1}^{K} P(r_i|s_k)P(s_k|d_j)} = \frac{P(r_i|s_k)P(s_k|d_j)}{\sum_{l=1}^{K} P(r_i|s_l)P(s_l|d_j)} \tag{9}$$

Then, refering to the likelihood principle, $P(d_j)$, $P(r_i|s_k)$, and $P(s_k|d_j)$ can be determined by maximizing the log-likelihood function:

$$L = \log\left(P(R,D,S)\right) = \sum_{i=1}^{N}\sum_{j=1}^{M} n\left(r_i, d_j\right) \sum_{k=1}^{K} P\left(s_k | r_i, d_j\right) \log\left[P\left(s_k | d_j\right) P\left(r_i | s_k\right)\right] \tag{10}$$

where $n(r_i,d_j)$ indicates the number of occurrences of region r_i in image d_j.

The standard procedure for maximum likelihood estimation is the EM algorithm. This method has two steps: expectation step (E-step) and maximization step (M-step). The E-step can be interpreted as mining the relationship between current estimates of the parameters and the latent variables by computing posterior probabilities. The M-step can be interpreted as updating parameters based on the so-called expected complete-data log-likelihood.

According to the EM method, the process of obtaining Equation (8) can be considered as the E-step, and the process of obtaining Equation (9) can be considered as the process of log-likelihood estimation. Then, Equation (9) is maximized using Lagrange multipliers. Equations (11) and (12) can then be derived

$$P\left(r_i | s_k\right) = \frac{\sum_{j=1}^{M} n\left(r_i, d_j\right) P\left(s_k | r_i, d_j\right)}{\sum_{m=1}^{N} \sum_{j=1}^{M} n\left(r_m, d_j\right) P\left(s_k | r_m, d_j\right)}, \tag{11}$$

$$P\left(s_k | d_j\right) = \frac{\sum_{i=1}^{N} n\left(r_i, d_j\right) P\left(s_k | r_i, d_j\right)}{\sum_{n=1}^{N} n\left(r_n, d_j\right)} \tag{12}$$

where n is the number of regions and region features, $n \in \{1,...,N\}$.

The E-step and M-step equations are calculated alternately until a local maximum of the expectation in Equation (9) is found. Because the distributions of $P(R|S)$, $P(S|D)$, and $P(S|R,D)$ are uniform, their initial values can be set equal to $P(R|S)$. The number of iterations depends on experience; in this research, it is set to five.

Each image can then be represented by the posterior probability $P(s_k|d_j)$ instead of by the original image feature.

EXPERIMENTS

In the experiments, TM images of Kii Peninsula (Japan), Wuhan (China), and Yancheng (China) are used. Each image is split into 256 × 256 subimages, and the total number of images is 2,000. Each image could manually be classified into eight land-cover types, namely sea, river, lake, farmland, urban area, cloud, forest, and bare soil.

Image Segmentation

Experiments on TM image are performed to test this segmentation algorithm. For a comprehensible comparison, we use original JSEG algorithm and the well-established eCognition. The results are shown in Figure 5 (boundaries are highlighted by red lines).

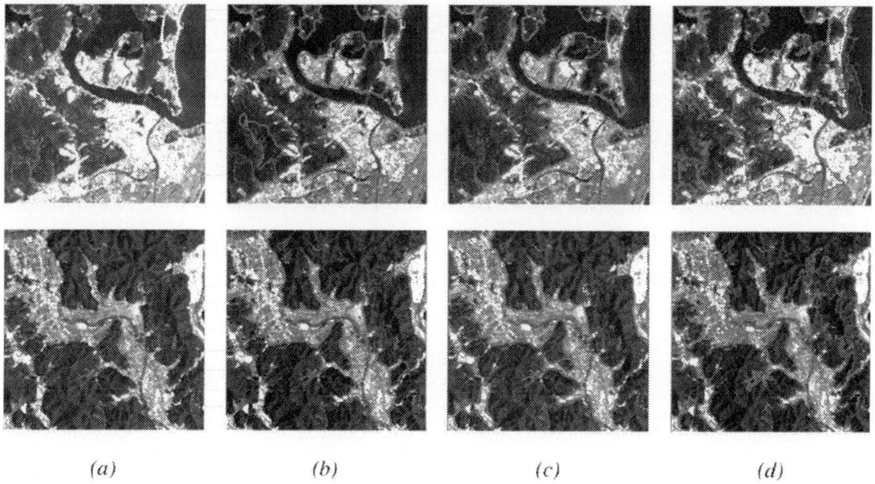

Figure 5: Images and corresponding segmentation results using different methods. (a) The original TM images with forest, urban area, and sea. (b-d) The results that are separately segmented by original JSEG algorithm, the proposed method, and eCognition, respectively.

Figure 5a shows two original TM images, both covering urban area and forest. The textural characteristics are clear in the forest area; because of the modest resolution of the TM sensor, spectral characteristics are more prominent than textural characteristics in the urban area. Figure5b presents the results from the original JSEG method. Note that the JSEG method produces good results, but sometimes it cannot separate two different regions very well due to the complex spectral characteristics of remotely sensed image. Compared with the JSEG method, the proposed method (Figure 5c), which takes NDVI, NDBI, and textural features into consideration, makes the difference between the ground covers more obvious and generates much better segmentation results. Figure 5d presents the results obtained from eCognition. Although the boundary is clear in Figure 5d, the result contains many fragments and some oversegmentation.

These experimental results prove that the proposed method outperforms the other two methods in terms of visual evaluation. It not only produces a good segmentation boundary, but also avoids oversegmentation.

Semantic Feature Extraction

In this experiment, semantic features are extracted from spectral and textural features. To determine the optimal number of semantic features, different numbers of semantic features are used to retrieve images. Without considering time requirements, the retrieval precisions obtained for the initial 20(40) result images (denoted as Top (20(40))) are shown in Figure 6.

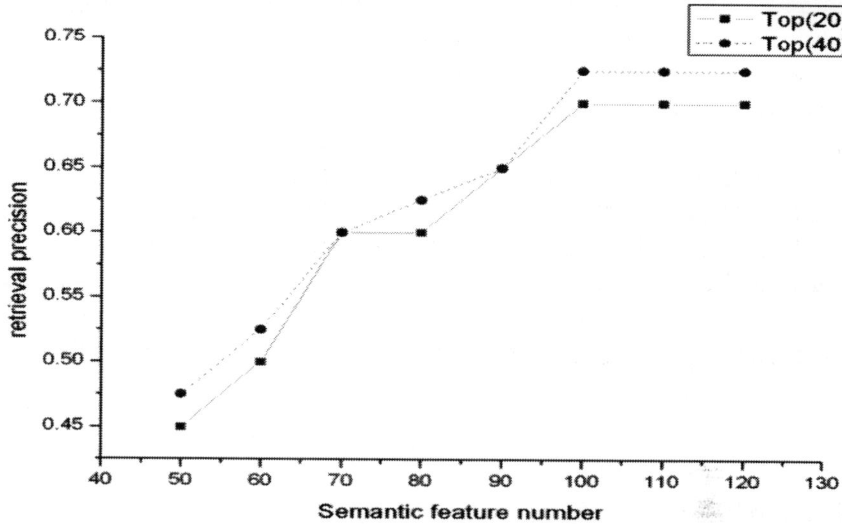

Figure 6: Retrieval precision for different numbers of semantic features.

The result indicates the general trend that the larger the number of semantic features, the higher the retrieval precision is. This occurs because more semantic features are used to describe the image content, more details can be described. However, a larger number of semantic features will lead to greater computational complexity for hidden semantic feature extraction and time requirements for computing similarity for image retrieval. The first turning point in Figure 6 is 100; there is no change when the number of semantic features is larger than 100. Therefore, the number of semantic features is set to 100.

Differences between Semantic Features

In this experiment, two groups of original remotely sensed image and their corresponding semantic features are shown in Figures 7 and 8, respectively.

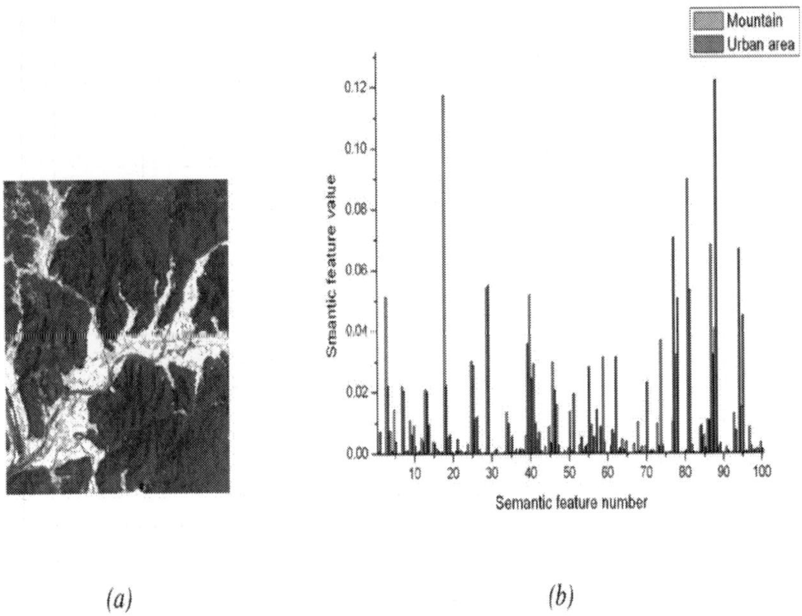

Figure 7: Different land cover types and corresponding semantic features. (a) Original image. (b) Semantic feature of mountain and urban areas.

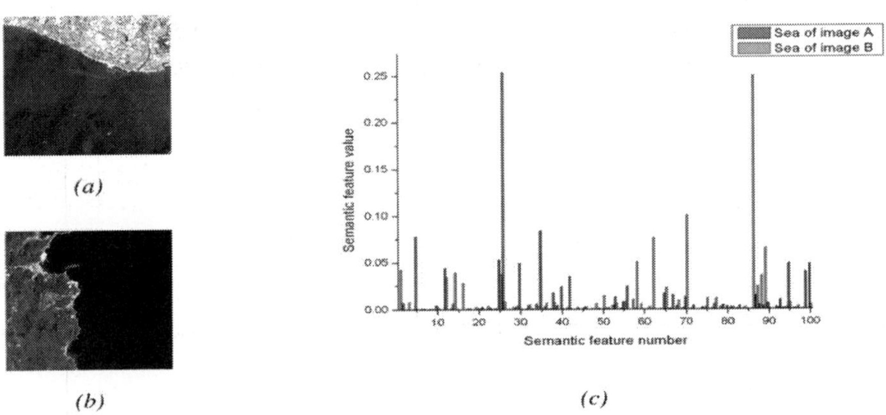

Figure 8: Similar land cover types and corresponding semantic features. (a) Original image. (b) Original image. (c) Semantic feature of sea of (a) and (b).

Figure 7a shows an image covering forest and urban area; Figure 7b shows the semantic features of these two kinds of ground cover. In Figure 7b, the cyan column presents semantic feature of mountain, while the red column presents semantic feature of urban area; each column indicates the value of corresponding semantic feature dimension. The higher the column, the larger the semantic feature value is. The total number of semantic features is 100.

In the original image (Figure 7a), it is clear that mountainous and urban areas present obviously different features; in Figure 7b, this difference is also clearly presented. For examples, in the 18th semantic feature, the cyan column is extremely higher than the red, while in the 88th semantic feature, the red column is extremely higher than the cyan. This indicates that semantic features can be used as a replacement for low-level features to distinguish between different ground covers.

Figure 8a, b presents two images both covering sea area. Figure 8c shows the semantic features of these two sea images. In Figure 8c, the red column presents semantic features of the sea area of image a, while the cyan column presents semantic features of the sea area of image b; each column indicates the value of the corresponding semantic feature dimension. The higher the column, the larger the semantic feature value is, and the total number of semantic features is 100.

Although the characteristics of the sea area in these two images are same, the concept is different because the area adjacent to sea in Figure 8a is urban while in Figure 8b it is mountain. This different concept between two sea areas is clearly presented in Figure 8c. For examples, in the 26th semantic feature, the red column is extremely higher than the cyan while in the 86th semantic feature, the cyan column is extremely higher than the red. This indicates that these semantic features can also well describe the high-level semantic concepts hidden in images.

These two experiments lead to the conclusion that the extracted semantic features can well describe not only low-level image characteristics, but also high-level hidden concepts.

Image Retrieval Experiments

Once the semantic features have been extracted, each image can be represented. The Euclidean distance method is used to calculate the similarity between two images. The following two experiments present the different specimen images and the top 20 retrieved results.

Experiment A

In this retrieval experiment, the specimen image is shown in Figure 9a, with the corresponding retrieval results consisting of the most similar images presented in Figure 9b. Specimen image covers mountain area and urban area, in which low-level features (spectral and textural) are clearly seen, and semantic perception is that urban area is on the foot of a mountain.

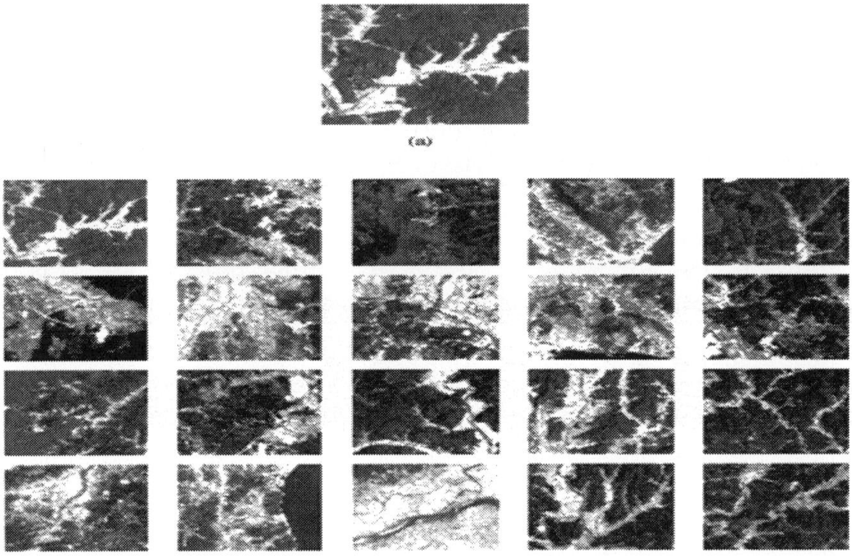

Figure 9: Image retrieval results. (a) Specimen image covers mountain and urban area. (b) The top 20 retrieval results.

The analysis of spectral and textural features shows that most of these 20 retrieval results are similar with the specimen image covering not only mountain, but also urban areas. The analysis of high-level semantic features shows that 14 images are similar with the specimen image in which urban area is surrounded by mountain, but others are different in which mountain and urban areas are adjacent.

Experiment B

In this experiment, an image covering cloud and mountain is chosen as the specimen image. The semantic perception in this image is that clouds float above mountain. The top 20 retrieval results, consisting of the most similar images, are shown in Figure 10b. The analysis of low-level features shows that most of results are similar to specimen image. However, according to semantic perception, 16 of them are similar to the specimen image.

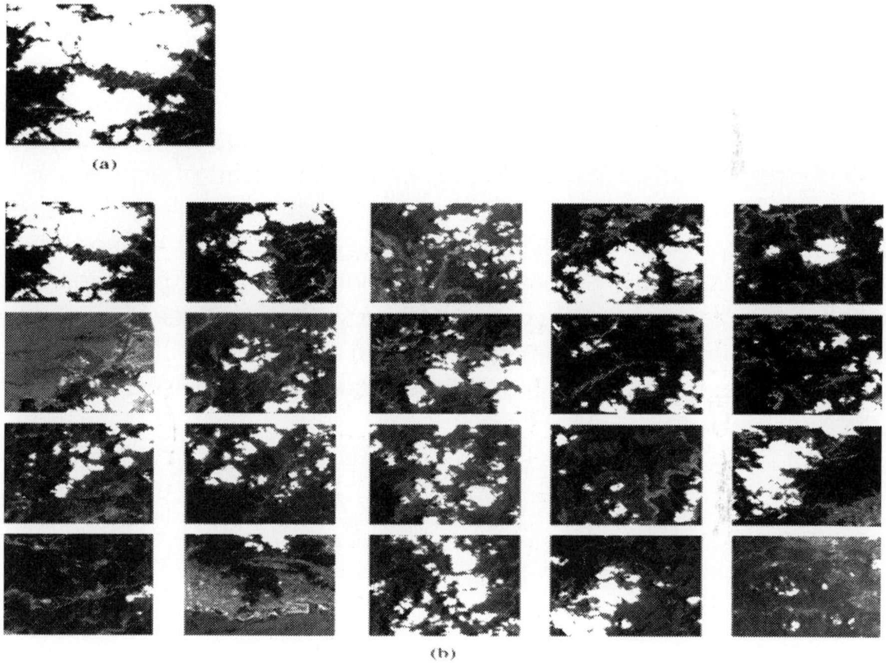

Figure 10: Image retrieval results. (a) Specimen image covers cloud and mountain. (b) The top 20 retrieval results.

Image Retrieval Precision and Recall

According to the results as shown in Figures 9 and 10, retrieval precision and recall are calculated. The precision and recall used in this analysis are defined as Equations (13) and (14)

$$\text{Precision} = \frac{I_{\text{relevant}} \cap I_{\text{retrieved}}}{I_{\text{retrieved}}} \tag{13}$$

$$\text{Recall} = \frac{I_{\text{relevant}} \cap I_{\text{retrieved}}}{I_{\text{relevant}}}, \tag{14}$$

where I_{relevant} is the total number of relevant images, and $I_{\text{retrieved}}$ is the total number of retrieved images. Figures 11 and 12 present the precision and recall results when the numbers of total retrieved images are 10, 20, 30, and 40, respectively. Although the precision and recall of experiment B are much higher than those of experiment A, both of them exhibit a slow drop in precision by increasing the number of retrieved images while the recall increases. It is consistent with normal trend of precision and recall. For a comprehensible comparison, low-level features of specimen images are used to retrieve image. Precision and recall results are shown in each figure. It can be noticed that the proposed method obtains higher precision and recall.

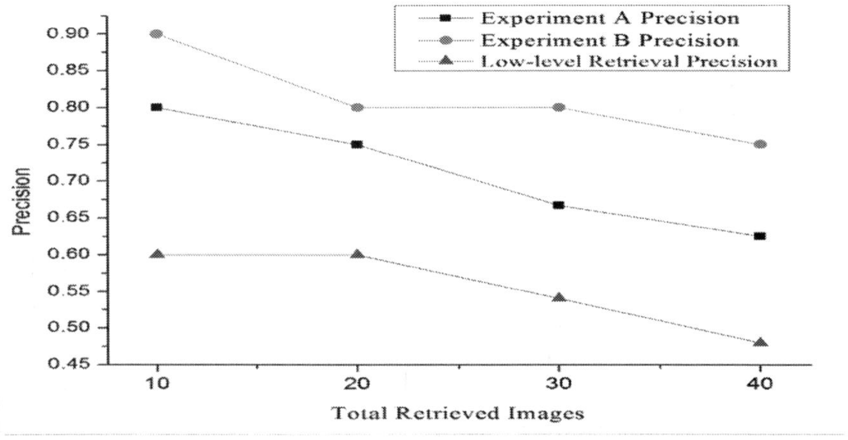

Figure 11: The precision results for different number of total retrieved images.

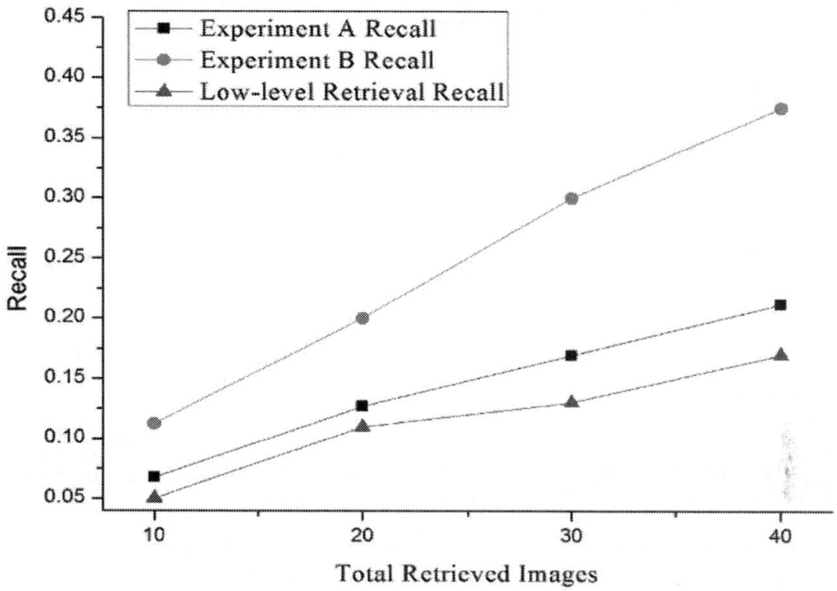

Figure 12: The recall results for different number of total retrieved images.

CONCLUSIONS

In this article, a region-level semantic-based satellite image retrieval system is described between low-level information and high-level concepts. Regional and semantic features are combined to narrow the semantic gap. An improved image segmentation algorithm is introduced which can obtain much better segmentation results than earlier algorithms. Semantic features are extracted using a probabilistic method, and experiments indicate that the new semantic features can represent not only low-level information, but also high-level concepts. Image retrieval experiments on two different specimens attain better retrieval precision and recall. The major limitation of this approach, however, is that a computationally expensive regeneration of the derived semantic model is required if new satellite images are added to the database.

ACKNOWLEDGEMENTS

This research was supported by the Chinese 863 program (No.2009AA12Z133), the National Nature Science Foundation of China (NSFC) (No. 41076126/D0611), and the National Basic Research Program of China (No. 2011CB7071).

REFERENCES

1. S Aksoy, K Koperski, C Tusk, G Marchisio, JC Tilton, Learning Bayesian classifiers for scene classification with a visual grammar. IEEE Trans Geosci Remote Sens 43(3), 581–589 (2005)
2. AD Bimbo, *Visual Information Retrieval* (Morgan Kaufmann Publishers, Inc., San Francisco, CA, 1999)
3. N Vasconcelos, From pixels to semantic spaces: advances in content-based image retrieval. Computer 40(7), 20–26 (2007)
4. M Flickner, H Sawhney, W Niblack, J Ashley, H Qian, B Dom, M Gorkani, J Hafner, D Lee, D Petkovic, D Steele, P Yanker, Query by image and video content: the QBIC system. Computer 28(9), 23–32 (1995).
5. JR Smith, SF Chang, Automated binary texture feature sets for image retrieval. *IEEE International Conference on Acoustics, Speech, and Signal Processing* (Atlanta, GA, USA, 1996) 4, pp. 2239–2242
6. YD Chun, NC Kim, Content-based image retrieval using multiresolution color and texture features. IEEE Trans Multimedia 10(6), 1073–1084 (2008)
7. M Ferecatu, N Boujemaa, Interactive remote-sensing image retrieval using active relevance feedback. IEEE Trans Geosci Remote Sens 45(4), 818–826 (2007)
8. X Huang, LP Zhang, PX Li, Classification and extraction of spatial features in urban areas using high resolution multispectral imagery. IEEE Trans Geosci Remote Sens Lett 4(2), 260–264 (2007)
9. C Diou, G Stephanopoulos, P Panagiotopoulos, C Papachristou, N Dimitriou, A Delopoulos, Large-scale concept detection in

multimedia data using small training sets and cross-domain concept fusion. IEEE Trans Circ Syst Video Technol 20(12), 1808–1821 (2010)

10. P Mylonas, E Spyrou, Y Avrithis, S Kollias, Using visual context and region semantics for high-level concept detection. IEEE Trans Multimedia 11(2), 229–243 (2009)
11. Y Li, TR Bretschneider, Semantic-sensitive satellite image retrieval. IEEE Trans Geosci Remote Sens 45(4), 853–860 (2007)
12. CS Lee, WY Ma, HJ Zhang, Information embedding based on user's relevance feedback for image retrieval. *Proc of SPIE Photonics East* (Boston, MA, 1999), pp. 19–22
13. Y Rui, TS Huang, M Ortega, S Mehrotra, Relevance feedback: a power tool for interactive content-based image retrieval. IEEE Trans Circ Syst Video Technol 8(5), 644–655 (1998).
14. J Li, RM Narayanan, Integrated spectral and spatial information mining in remote sensing imagery. IEEE Trans Geosci Remote Sens 42(3), 673–685 (2004).
15. QY Li, H Hu, ZZ Shi, Semantic feature extraction using genetic programming in image retrieval. *Proceedings of the 17th International Conference on Pattern Recognition (ICPR)* (Cambridge, England, UK, 2004) 1, pp. 648–651
16. M Datcu, H Daschiel, A Pelizzari, M Quartulli, A Galoppo, A Colapicchioni, M Pastori, K Seidel, PG Marchetti, S D'Elia, Information mining in remote sensing image archives: system concepts. IEEE Trans Geosci Remote Sens 41(12), 2923–2936 (2003).
17. H Daschiel, M Datcu, Design and evaluation of human-machine communication for image information mining. IEEE Trans Multimedia 7(6), 1036–1046 (2005)
18. YN Deng, BS Manjunath, Unsupervised segmentation of color-texture regions in images and video. IEEE Trans Pattern Anal Mach Intell 23(8), 800–810 (2001).
19. ZY Wang, R Boesch, Color- and texture-based image segmentation for improved forest delineation. IEEE Trans Geosci Remote Sens 45(10), 3055–3062 (2007)
20. Y Chang, D Lee, Y Wang, Color-texture segmentation of medical images based on local contrast information. *Proc IEEE*

Symposium on Computational Intelligence and Bioinformatics and Computational Biology (Honolulu, HI, 2007), pp. 488–493

21. JW Han, KN Ngan, Automatic segmentation of objects of interest in video: a unified framework. *Proceeding of 2004 Intelligent Signal Processing and Communication Systems (ISPACS)* (Seoul, South Korea, 2004), pp. 375–378
22. A Gersho, RM Gray, *Vector Quantization and Signal Compression* (Kluwer Academic: Norwell, MA, 1992)
23. F Dell'Acqua, P Gamba, A Ferrari, JA Palmason, JA Benediktsson, K Arnason, Exploiting spectral and spatial information in hyperspectral urban data with high resolution. IEEE Geosci Remote Sens Lett 1(4), 322–326 (2004).
24. JW Rouse, RH Haas, JA Schell, DW Deering, in *Monitoring vegetation systems in the great plains with ERTS*, ed. by Freden SC, Mercanti EP, Becker MA (Third Earth Resources Technology Satellite-1 Symposium--Volume I: Technical Presentations, NASA SP-351, Washington, USA, 1974), pp. 310–317
25. Y Zha, J Gao, S Ni, Use of normalized difference built-up index in automatically mapping urban areas from TM imagery. Int J Remote Sens 24(3), 583–594 (2003).
26. YD Zhao, LP Zhang, PX Li, Texture feature fusion for high resolution satellite image classification. *International Conference on Computer Graphics, Imaging and Vision: New Trends* (Beijing, China, 2005), pp. 19–23
27. RM Haralick, Statistical and structural approaches to texture. Proc IEEE 67(5), 786–804 (1979)
28. AP Dempster, NM Laird, DB Rubin, Maximum likelihood from incomplete data via the EM algorithm. J R Stat Soc Ser B 39(1), 1–38 (1977)
29. TK Moon, The expectation maximization algorithm. IEEE Signal Process Mag 13, 47–60 (1996

Chapter 7

A BIM-Based Monitoring System for Urban Deep Excavation Projects

I-Chen Wu, Siang-Rou Lu, and Bin-Chen Hsiung

Department of Civil Engineering, National Kaohsiung University of Applied Sciences, Kaohsiung, Taiwan

ABSTRACT

Background

Deep excavations in urban areas have the potential to cause unfavorable effects on ground stability and nearby structures. Thus, it is necessary to evaluate and monitor the environmental impact during deep excavation construction processes. Generally, construction project teams will set up monitoring instruments to control and monitor the overall environmental status, especially during the construction of retaining walls, main excavations, and when groundwater is involved. Large volumes of monitoring data and project information are typically created as the construction project progresses, making it increasingly difficult to manage them comprehensively.

Methods

To address the abovementioned issues, this project develops a Building Information Model (BIM)-based monitoring system to integrate and visualize monitoring data for risk assessments during urban deep excavation projects. A BIM can be used to establish a comprehensive model of managing a construction project. The system developed herein can access required data from BIM models, and allow complicated numerical data to be displayed effectively in an easily understandable visual format. It is composed of intelligent building components, which includes data attributes and parametric rules for each object. This system can provide a construction project team with a full monitoring view of the ongoing project, along with functions to integrate the information and display it in various ways to present complex engineering monitoring information quickly and clearly.

Results

Testing examples of the developed system on the excavation of the O6 station in the Kaohsiung metro system in Taiwan are presented to illustrate the improvements of safety management for adjacent structures in urban deep excavation projects. The risks and issues affecting the safety of excavation activities and proximal structures during a project can be identified earlier through effective visualization of information in the system, enabling construction project teams to address them promptly and appropriately by performing accurate risk assessments and decision making.

Conclusions

This system assists construction project teams in identifying and understanding possible blind spots when attempting to achieve risk assessments during urban deep excavation projects, and further enables the adoption of mitigation measures to reduce risk levels.

BACKGROUND

As a result of rapid urbanization, underground structures such as deep excavations and tunnels; have been widely used to increase the availability of space by moving underground. As these structures are often constructed below mature and densely populated city areas, any associated accidents can easily result in serious damage to adjacent structures or cause human casualties. Ground settlement is especially thought to have the greatest impact on the nearby environment. At present, numerical analysis methods and field monitoring are two commonly used tools to predict and assess the ground settlement induced by an excavation.

Numerical Analysis Method

Through the collection and evaluation of data of surface settlement from underground construction in various locations, Peck ([1976]) first defined the possible zone of influence around an excavation. Bowles ([1988]) recommended a procedure for estimating the excavation-induced ground surface but required the lateral wall deflection to first be calculated. Clough and O'Rourke ([1990]) proposed various types of envelopes of excavation-induced ground surface settlements and these were mainly classified by ground conditions. Ou ([2006]) developed a method to predict the details of ground surface settlement on the basis of studies, and concluded that in principle there are two types of settlement curves, the spandrel and the concave; and that the magnitude of wall deflection in a cantilever model would define the type of settlement curve. Similarly, various studies have been conducted on the application of numerical analyses in deep excavations, but they are not addressed here owing to limits on paper length.

Field Monitoring

The influence of adjacent structures can only be identified with the effective use of monitoring data obtained from field instruments.

Professional engineering knowledge and experience can be used to interpret the above-mentioned results to avoid and reduce accidents from occurring. Related studies, such as that by Qiu et al. ([2009]) described a monitoring and analysis information system to analyze the data of construction monitoring and provide timely information feedback for safety in urban subway construction. Shao and Macari ([2008]) presented a systematic procedure, referred to as "Information Feedback Analysis", which is used to predict excavation-induced deformations by collecting field information such as displacement. Ran et al. ([2011]) designed and implemented a long-term monitoring and safety evaluation system for the deep excavation of a metro station. However, the system behavior could only be identified with the effective use of monitoring data obtained from field instruments.

Data Visualization and Communication

Here this project limits the discussion to data management of field monitoring. It is well known that the design and construction of deep excavations increase in difficulty as the depth of an excavation increases and the surrounding environment increases in complexity. Therefore, a construction project team must consider a wide variety of information to assess the environmental impact when managing risks and making project decisions. Unfortunately, engineers understand field monitoring data has been limited and such understanding becomes increasingly difficult to acquire with the growing amount of data. Treicher ([1967]), an experimental psychologist, proved that of the information human beings receive, 83% is by sight, and this shows that information visualization is essential for communication and information distribution. With the growing use of visualization techniques in construction, Building Information Modeling (BIM) and Geographic Information Systems (GIS) have recently become widely used in the visualization of construction progress (Elbeltagi and Dawood,[2011]).

2D Data Visualization

As an example of a GIS application, Zhou et al. ([2012]) developed a natural environment information management system. Its core concept lies in collecting effective and reliable environmental information, increasing utilization rates, sharing a degree of environmental information by advanced information technology, and providing timely and scientific foundations for environmental monitoring and management. Li and Li ([2012]) designed a monitoring and warning system for facilitating the reduction of capital investment and preventing geological disasters. The automation and information processing for landslide hazards in this system can provide a basis for early warning of landslides in a village called Diao Zhongba. Şalap et al. ([2009]) presented the development of a GIS-based monitoring and management system for underground mine safety in three levels, namely constructive safety, surveillance and maintenance, and emergency. The proposed system is expected to be an efficient tool for improving and maintaining healthy standards in underground coal mines and has the potential to be extended to a national GIS infrastructure. GIS facilitates monitoring underground safety by allowing the use of spatially referenced data, in which the system functions through its database and management tools, processing and analysis tools, and featured display tools (Dijk, PM van et al. [2004]). However, a GIS application mainly focuses on a 2D plan, and thus cannot provide an accurate source of information about the height and depth in a 3D space.

3D Data Visualization

Meanwhile, BIM is a relatively new technology that facilitates better information integration and management (Eastman et al. [2008]). Many engineering companies employ BIM for information integration, visualization, and parametric design to reduce both the duplication of work and the complexity of interface integration; and to assist construction project teams in managing risks and making project decisions. BIM offers detailed 3D visualization and the ability to organize large volumes of data related to buildings for management. Hsieh and Lu ([2012]) demonstrated that a visualization system enables users to interpret monitoring data effectively and intuitively, reduces misinterpretation, improves communication efficiency, and facilitates efficient decisions that are supported by the monitoring data gathered. Kim et al. ([2012]) focused on the monitoring and visualization of aggregated and real time states of various energy usages represented by location-based sensor data accrued from city to building scale. Design representations are no longer 2D drawings. Instead, designers use 3D BIM models that are assembled in the same way a building is constructed. BIM can be used in construction firms to better reuse the accumulated management information. Employing information within the BIM model allows everyone on the project team to make better and more-informed decisions across the entire project lifecycle of building and infrastructure projects.

Project Aim and Objective

For above mentioned reasons, this project took advantage of BIM with regard to integration and visualization to develop a BIM-based monitoring system that aims to integrate the required data and to visualize this data appropriately and effectively. It is expected that this system will assist construction project teams to effectively assess risks and manage monitoring data. The concept of this project is shown in Figure 1.

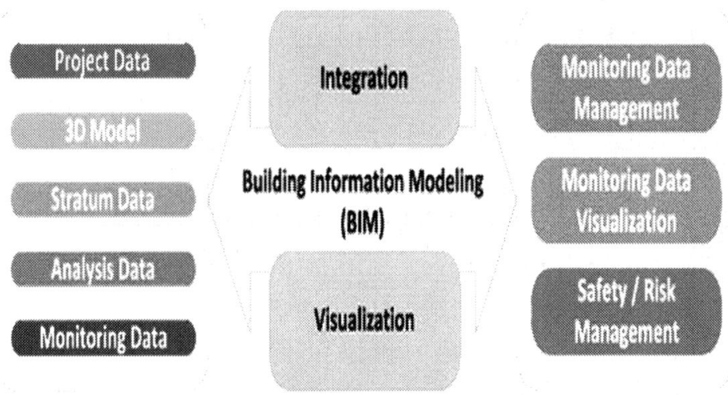

Figure 1: The concept of this project.

METHODS

A BIM-based monitoring system was developed in this project. The system is implemented based on commercial hardware and software comprising the Bentley AECOsim Building Designer, which supports visualization of a 3D model with some capabilities for 3D object manipulation and information query with Application Programming Interfaces (API) for functionality extensions. A Mass Rapid Transit (MRT) deep excavation project was chosen as a case study for demonstrating the results of this project. Finally, environmental and monitoring data were integrated and visualized to allow users to quickly obtain the information required to make decisions and conduct evaluations to solve problems and plan appropriate emergency measures. A detailed discussion is presented in the next section.

System Framework

Figure 2 depicts the basic framework of the system design. The framework consists of two major parts, the Kernel and Graphic User Interface (GUI) subsystems. The Kernel subsystem defines a higher level and unified interface to the AECOsim BuildingDesigner DGN 2.0 Library (MicroStationDGN) and is divided into the following three modules: (1) DataManagement, which includes many different classes to enable the provision of functions for manipulating objects such as I/O, modify, delete, and query; (2) DataAnalysis, which provides indispensable calculation functions of risk assessment and quality statistics; and (3) DataVisualization, which drives MicroStationDGN's visualization engine to provide some visualization styles for communication between the project participants. The GUI forms are available for a user to control and manipulate the BIM-based monitoring system.

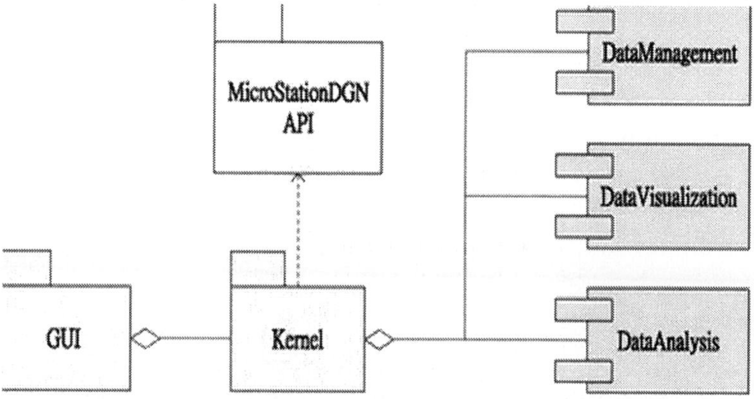

Figure 2: System framework of the BIM-based monitoring system.

Building Information Model

The Building Information Model is a computer model database of building design information, which may also contain information about the building's construction, management, operations, and maintenance (Howell and Batcheler, [2005]). In this project, the building information model becomes a shared knowledge resource to support risk assessment. The entire integrated dataset will be stored into a database (non-geometry) and an AECOsim dgn file (geometry). The sources of data for integration into the building information model in this project (as shown in Figure 3) are as follows:

Project Data

Project data is required for project participants to understand background information of the project, such as the project name, address, owner, and the construction company.

3D Model

BIM technology currently focuses on the visualization of the changing status of 3D shapes during the construction process. Therefore, this project integrates different types of 3D models into the BIM system and includes the building model, excavation model, retaining wall model, and supporting elements for visualization.

Stratum Data

The excavation performance differs according to different ground conditions. This system integrates soil properties and groundwater statuses for assisting construction project teams to understand the environmental status for making decisions.

Analysis Data

Generally, construction project teams will analyze and predict ground settlement and the influence zone using empirical methods prior to a deep excavation. In this project, these results were integrated and visualized in the BIM system to assist construction project teams to assess the environmental impact in a synthesized manner.

Monitoring Data

Construction project teams normally set up various types of monitoring instruments to collect the necessary data to monitor the status of different environmental variables. The related data integrated into the BIM system include types, standard values, and units recorded by the various field-monitoring instruments.

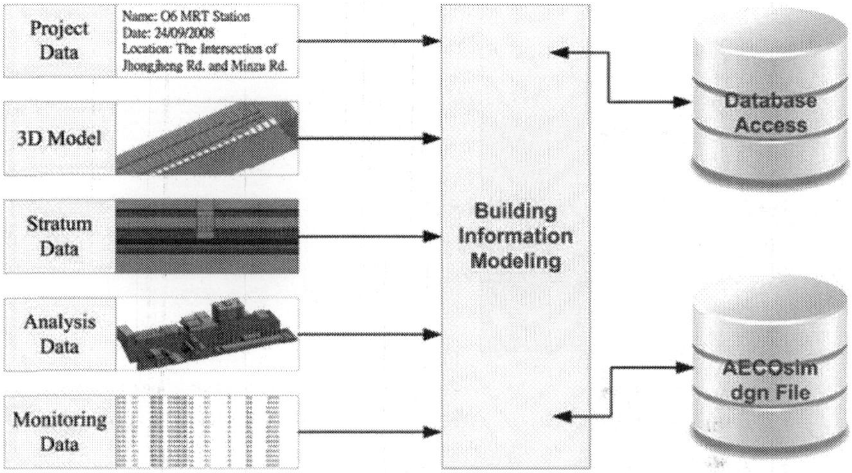

Figure 3: The sources of data for integration in the BIM.

Monitoring data management

The goal of monitoring is to provide useful scientific information about the status and trends of various factors affecting the environmental impact of a project. Comprehensive data management functionality is thus essential for achieving this goal, particularly for urban deep excavations.

Data Input

This system provides two methods for data input, as shown in Figure 4. For visualizing the location of monitoring instruments accurately in the BIM system, the different kinds of monitoring instruments are represented by the different 3D primitive solids shown in representation of a monitoring instrument can be created from the system and then the field data of the monitoring instrument can be entered or updated individually. Moreover, as monitoring instruments are updated frequently, the system needs to provide a batch input function to effectively deal with the large amount of monitoring data arriving from the database at any one time.

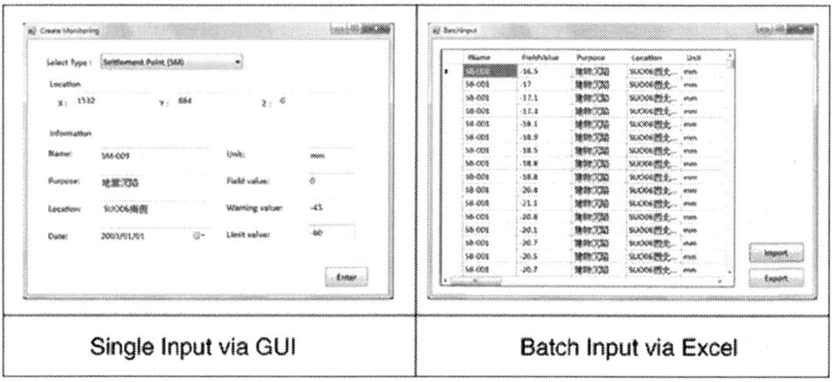

Figure 4: The two methods for data input.

Data Access

As shown in Figure 5, this project designed and developed an integrated a BIM model to describe and store information including 3D models, stratum data, analysis data, project data, and monitoring data for an entire project. All project data can be divided into two types: non-graphic data, which is stored into a database; and a 3D geometry, which is stored in the AECOsim object model (a .dgn file). Each object (graphic element) that is linked to a database has a database linkage attached to the element. This is sometimes called the MSLINK in Bentley Systems. Users can double-click on the 3D object in AECOsim and summon detailed information on the 3D object that was extracted from the database and vice versa, as shown in Figure 6.

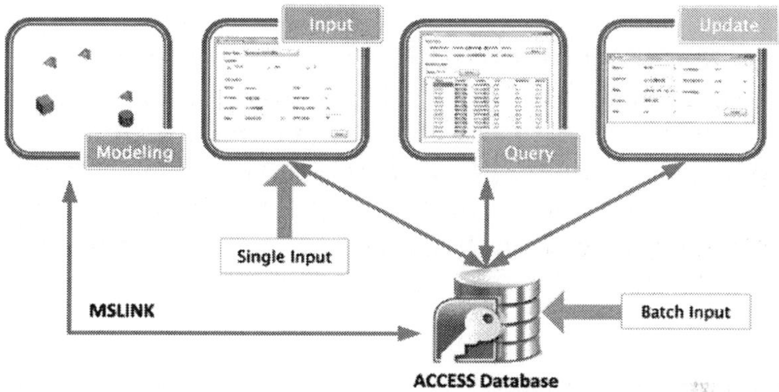

Figure 5: The concept of data access.

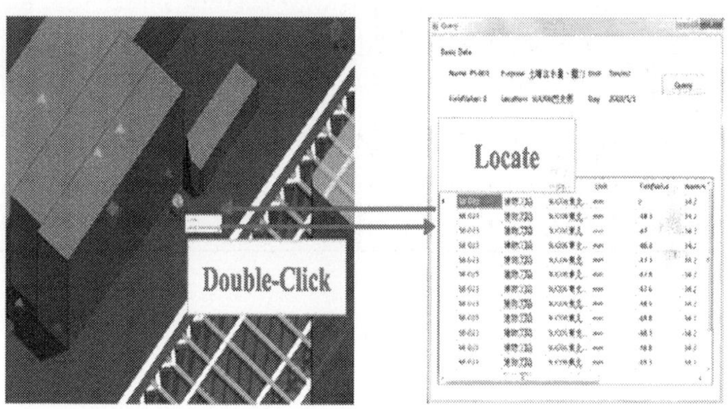

Figure 6: Data query between a 3D environment and the GUI.

Monitoring Data Visualization

Effective visualization assists people in obtaining the required information effectively and efficiently. This system was designed to not only provide sufficient information to facilitate project management, but also provide various visualization tools to assist with risk assessment and communication. This system provides the following three styles of visualization for communication between the project participants: (1) 1D (text and numeric values); (2) 2D (graphs and charts); and (3) 3D (3D models).

1D & 2D Visualizations

As shown in Figure 7, 1D Visualization produces text and numeric values for displaying attributes of objects, general project information, and monitoring data; while 2D Visualization shows graphs and charts for presenting the results of calculations and analyses between objects. A user can select a monitoring instrument for query and the system will show its information and status in detail.

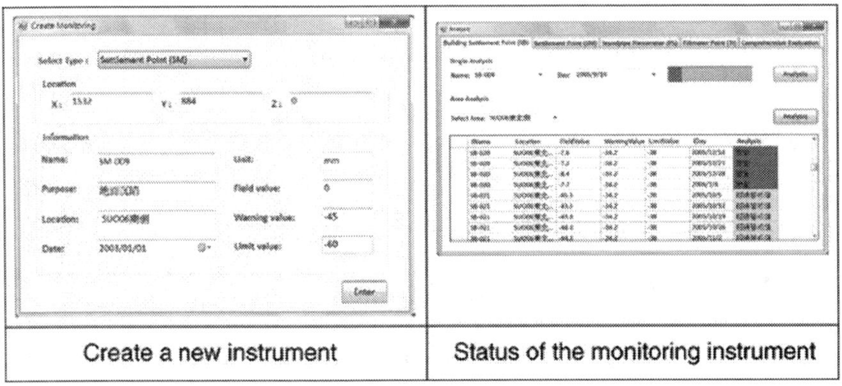

Figure 7: The 1D and 2D Visualization styles.

3D Visualization

As shown in Figure 8, 3D Visualization shows 3D models of buildings for presenting the actual status. In a 3D environment, this system resembles a virtual comprehensive central control room; it not only provides 3D manipulation functions such as fly, walk, and rotate, with no blind spots from the point of view of a user, but also controls all statuses of monitoring instruments in a construction site.

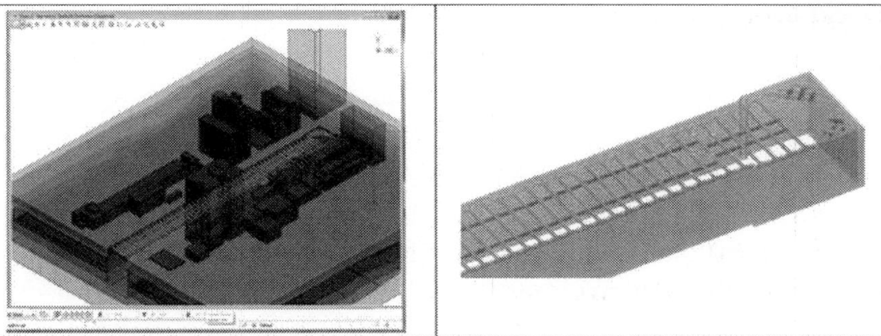

3D model of a construction site	3D model of a retaining wall.

Figure 8: The 3D Visualization style.

Safety/Risk Assessment

Urban deep excavations in cities quite often cause ground surface settlement in the city and main excavation activities. Therefore, this project mainly evaluates possible risks from the impact of these activities. Ground surface settlement can occur during three construction stages: (1) retaining wall construction, (2) groundwater pumping, and (3) main excavation. Therefore, project teams need to detect or predict the environmental impacts of these stages to prevent accidents from occurring. Recently, empirical methods and field monitoring techniques have become two commonly utilized tools to predict and assess the influence of adjacent structures by excavation.

Retaining Wall Construction

According to the monitoring results of the rapid transit system in Hong Kong, after the completion of the diaphragm walls and before the main excavation, the accumulated deformation was found to be 40–50% of the total deformation after the completion of the main excavation (Clough and O'Rourke,[1990]). Clough and O'Rourke ([1990]) found that the ratio of the maximum settlement induced by the construction of diaphragm walls to the depth of the trench was 0.15% according to in situmonitoring results.

Groundwater Pumping

Most problems encountered in deep excavations have a direct or indirect relationship with groundwater. Groundwater-induced problems in an excavation may arise from insufficient investigation of groundwater or geological conditions. Thus, it is necessary to perform detailed investigations of groundwater and its influences on soil or structures during excavations (Ou, [2006]).

Excavation

From observations of the shapes or types of ground surface settlement, it can be observed that the soil at the back of a retaining wall moves forward and down as the retaining wall deforms under normal conditions, producing ground settlement. Peck ([1976]) proposed that the zone of influence of a settlement should be two or three times that of the excavation depth. Clough and O'Rourke ([1990]) proposed that excavations in sandy soils might induce a zone of influence of settlement about twice that of the excavation depth (Ou, [2006]).

Empirical Methods

These ground conditions are important references for selecting a suitable excavation method, as adopting appropriate excavation methods can reduce risks. Stratum data were generated from a geological survey report. However, the report only presents the depth of soil for each drill hole. Therefore, this project integrates the depth of each drill hole and digital terrain model (DTM) to a 3D stratum data model, which is integrated and visualized in the BIM system, as shown in Figure 9(a). Moreover, different ground conditions and construction stages will require the application of different empirical methods to calculate the zone of influence of deep excavations. The BIM system can visualize the zone of influence of deep excavations (as developed by the previous researchers), as shown in Figure 9(b). The method for integrating these empirical formulas into the BIM system to assist construction project teams in evaluating construction statuses through various visualization tools is also shown. In addition, the system highlights the adjacent structures within the influence zone and presents details for a user to take action during early stages.

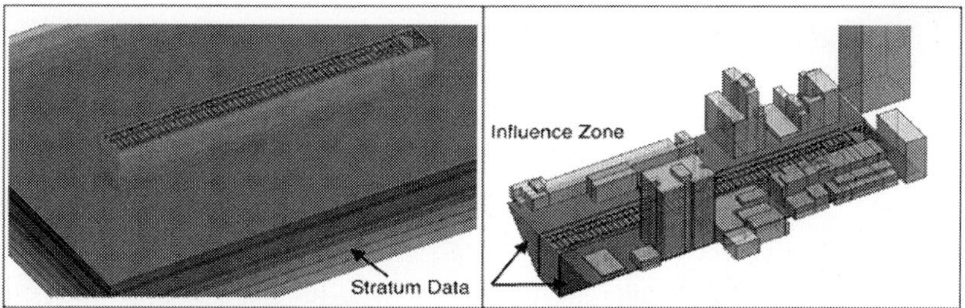

(a) Stratum data for environmental impact assessment	(b) Zone of influence of ground surface settlement

Figure 9: Visualizations for the numerical analysis methods.

Field Monitoring

Adequate risk management during construction is dependent on accurate and on-time reporting of key measures from the intensive monitoring system. The system can evaluate the probable environmental impacts from the critical monitoring data and the building information model and can enable construction project teams to identify and manage risk. Three colors are defined to visualize the different risk statuses in a 3D environment. The different kinds of monitoring instruments are represented by the different 3D primitive solids. shows the color scheme implemented for displaying each risk status. Furthermore, the 3D objects are highlighted in different colors according to their risk status. For example, if the field value of an instrument is more than the limit value, the representative 3D primitive solid will be showed in red, and the construction project team must urgently adopt mitigation measures to reduce the risk level.

The construction project team can input monitoring data from the GUI or the database. The system will subsequently update the representation of the status of environmental impacts by comparing field values and standard values, as shown in Figure 10(a). The construction project team can also view the overall status of the excavation project, as shown in Figure 10(b). The system can provide essential information for assisting planners in assessing environmental impacts comprehensively for safety and risk management.

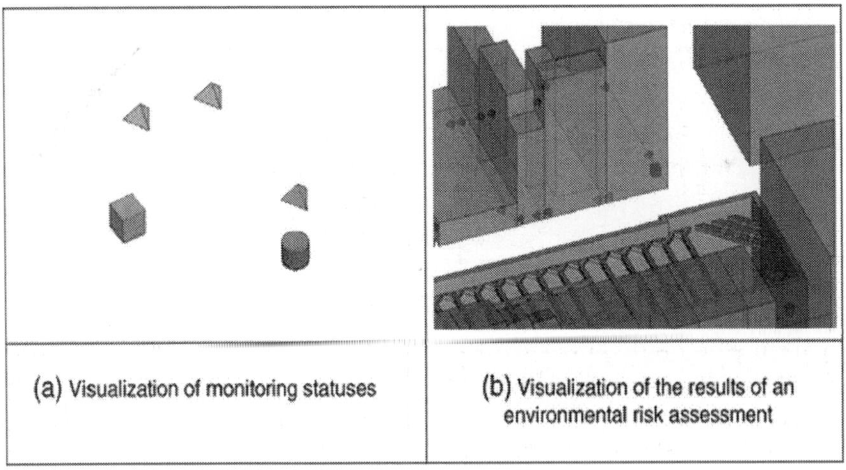

Figure 10: Visualization for field monitoring.

RESULTS AND DISCUSSION

Excavation of the O6 Station in the Kaohsiung metro system was selected as a case study for this project. Kaohsiung is the economic and political center of southern Taiwan. The metro project of the city commenced in 2001 and the O6 Station is located at the junction of Chung-Cheng Road and Ming-Chu Road. The surrounding area of the O6 Station is populous, as shown in Figure 11. The O6 Station was constructed by a cut-and-cover method and the maximum excavation depth was 19.6 m. The pit was retained by a 1-m-thick and 36-m-deep reinforcement concrete diaphragm wall. The major structure of the O6 Station is a 2-level basement. The main soil stratum of the site consists of a groundwater level, which was observed at 3.5 m below the surface level. Details of the project are given by Hsiung ([2009]). In such a complex environment, a construction project team must pay greater attention to the environmental impacts during a deep excavation project. Instruments installed on site include inclinometers in soils and diaphragm walls, settlement points, tiltmeters, and piezometers.This project uses this case for discussion and make comparisons between the current method and the proposed method in this project.

Data Management

Monitoring provides tangible information on a regular basis over an extended period of time about past and present conditions of the environment and thus, generates large amounts of data. Until now, many construction project teams have collected and analyzed monitoring data for each instrument manually, consequently wasting a lot of time. Moreover, these data will be stored in a file format. This approach is difficult for handling monitoring data used for querying and reviewing. In this project, all monitoring data will be collected and stored into a database and visualized based on a BIM system. A construction project team can manage monitoring data via a database or a visualization environment. These systems provide many functions for assisting a user to manage data efficiently and effectively.

Read and Discriminate

After collecting monitoring data, a construction project team needs to read and discriminate the data to determine the monitoring status for making decisions. Currently, analysis of monitoring data is conducted manually, and is error-prone. In this project, functions of risk assessment were implemented into the system to automatically facilitate reading and discriminating. This approach assists in reducing computational time and errors in the risk assessment processes.

Communication

Liston et al. ([2000], [2001]) studied the problem of communication in engineering workspaces, which could be variably defined as physical or virtual spaces where people work, share, and use information. From observations of numerous designs and construction review meetings, they discovered that teams spent most of their time on descriptive, explanative, evaluative, and predictive tasks. All such tasks are critical to enable improved decision-making by project managers. In this project, the developed system provides three styles of visualization for communication among the project participants; namely 1D (text), 2D (graph and chart), and 3D (3D Model). Figure 12 shows the overall monitoring data in a 3D visualization environment. A construction project team can control the construction status in single view. Figure 13 shows the results of a comprehensive evaluation via text and graphs. These visualization styles can conveniently provide different and sufficient information for different requirements for communication.

Figure 11: The main excavation zone of the MRT station in Kaohsiung.

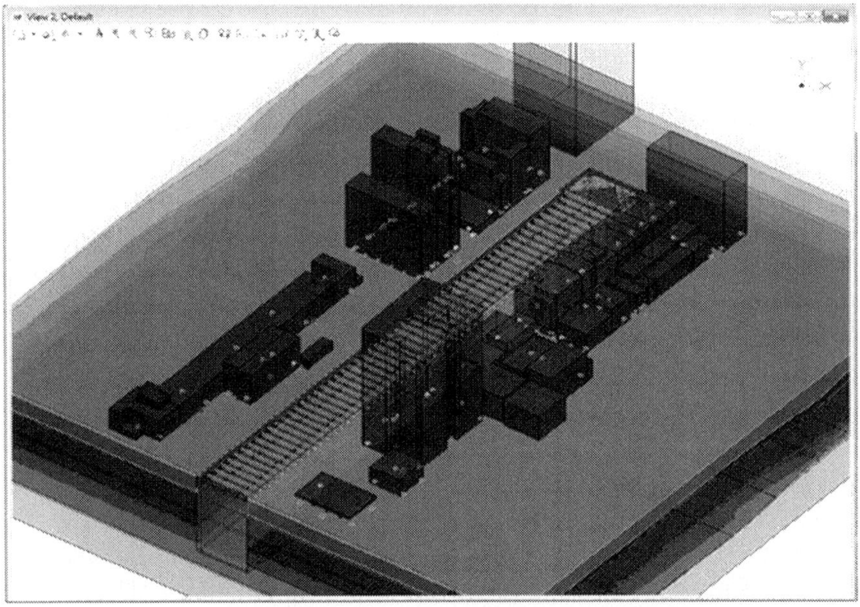

Figure 12: The monitoring overview.

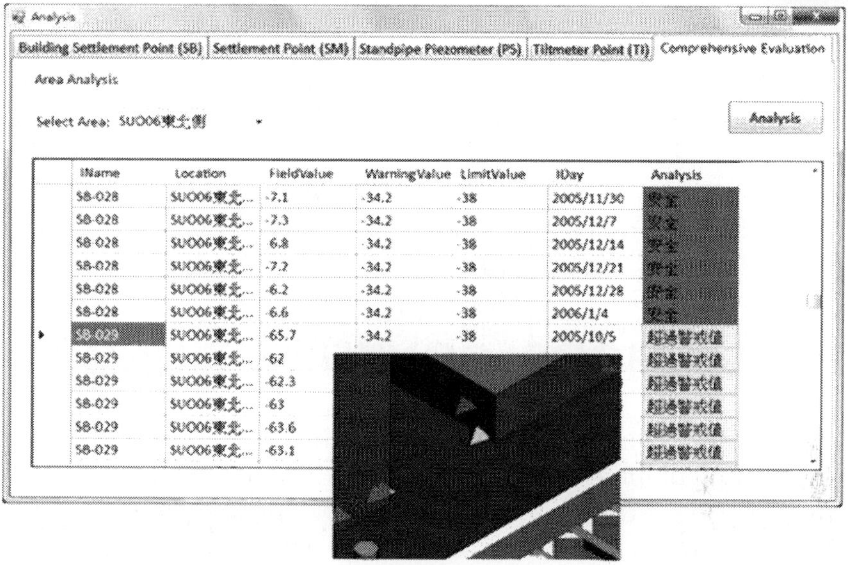

Figure 13: The results of a comprehensive evaluation.

CONCLUSIONS

Risk assessment is an important task for urban deep excavation projects. This project took advantage of the information integration and visualization capabilities of BIM to assist construction project teams with assessing possible environmental risks. The BIM-based monitoring system can provide users with a full overview of the ongoing urban deep excavation project, along with functions to integrate the information and display it in a multi-dimensional view. There are three main positive characteristics of the system. First, with an easy-to-use GUI, the efficiency of BIM manipulation and information acquisition is increased. Building information and monitoring data are also robustly stored and managed throughout all stages of an urban deep excavation project. Second, by representing the monitoring data and the results of an environmental risk assessment visually, the project team can quickly grasp the current state of an ongoing project, and communicate and coordinate accordingly with other project participants more efficiently. Third, through the comprehensive evaluation functionality, the project team can accurately monitor the project status and risks. This study demonstrated the use of a BIM of a deep excavation project to integrate a 3D model and relevant information about the retaining walls, excavation, and adjacent buildings; and then visualized all analysis and assessment results to present the likely locations and degrees of risk and safety under different situations. This was done to enable users to interpret monitoring data effectively and intuitively, reduce misinterpretation, improve communication efficiency, and facilitate efficient decisions that are supported by the gathered monitoring data.

AUTHORS' CONTRIBUTIONS

ICW and SRL developed the system together. All authors contributed to the writing of the manuscript. ICW and SRL drafted the manuscript and BCH reviewed and revised it. All authors read and approved the final manuscript.

REFERENCES

1. Bowles JE: Foundation Analysis and Design. 4th edition. McGraw-Hill, New York, USA; 1988.
2. Clough GW, O'Rourke TD: Construction-induced movements of in situ walls.Design and Performance of Earth Retaining Structure, ASCE Special Publications 1990, 25:439-470.
3. Dijk, PM van, Wang, HY, Genderen, JL van, (2004). Earth observation knowledge transfer: the example of ITC's coal fire project. Proceedings of the 20th International Society for Photogrammetry and Remote Sensing (ISPRS) Congress, Istanbul, Turkey.
4. Eastman C, Teicholz P, Sacks R, Liston K: BIM Handbook: A Guide to Building Information Modeling for Owners, Managers, Designers, Engineers and Contractors. Wiley, USA; 2008.
5. Elbeltagi E, Dawood M: Integrated visualized time control system for repetitive construction projects.Automation in Construction 2011, 20(7):940-953.
6. Howell, I, & Batcheler, B. (2005). Building Information Modeling Two Years Later - Huge Potential, Some Success and Several Limitations. The Laiserin Letter, May

7. Hsieh YM, Lu YS: Visualization system for field monitoring data and its effectiveness.Automation in Construction 2012, 26:54-68.
8. Hsiung BC: A case study on the behaviour of a deep excavation in sand.Computer and Geotechnics 2009, 36(4):665-675.
9. Kim SA, Shin D, Seibert T, Walz SP: Integrated energy monitoring and visualization system for Smart Green City development: Designing a spatial information integrated energy monitoring model in the context of massive data management on a web based platform.Automation in Construction 2012, 22:51-59.
10. Li, XP & Li, YA. (2012). Design of GIS-based Monitoring and Early-warning System of Landslide Hazard in Diao Zhongba. Energy Procedia, 16(B), 1174–1179.
11. Liston, K, Fischer, M, & Kunz, J. (2000). Designing and evaluating visualization techniques for construction planning, Proceedings of the 8th International Conference on Computing in Civil and Building Engineering, Stanford University, CA, USA, 1293–1300.
12. Liston K, Fischer M, Kunz J: Requirements and Benefits of Interactive Information Workspaces in Construction, CIFE Technical Report, Stanford University. 2001.
13. Ou CY: Deep Excavation: Theory and Practice. Taylor & Francis, Netherlands; 2006.
14. Peck RB:Deep excavations and tunneling in soft ground. Proceedings of the 7th International Conference on soil mechanics and foundation engineering, Mexico City. 1976.
15. Qiu D, Liang Q, Li Q, Wu R:Safety monitoring information system of urban subway.Proceedings of the 2009 WASE International Conference on Information Engineering, July 10–11, Taiyuan, Shanxi, China. 2009.
16. Ran L, Ye X, Zhu H: Long-term monitoring and safety evaluation of a metro station during deep excavation.Procedia Engineering 2011, 14:785-792.
17. Şalap S, Karslıoğlu MO, Demirel N: Development of a GIS-based monitoring and management system for underground coal mining safety.International Journal of Coal Geology 2009, 80:105-112.

18. Shao Y, Macari E: Information feedback analysis in deep excavations.International Journal of Geomechanics 2008, 8:91-103.
19. Treicher DG: Are you missing the boat in training aid?Film and AV Communication 1967, 1:14-16.
20. Zhou Q, Yuan Y, Kang Y: GIS application system design applied to information monitoring.Physics Procedia 2012, 25:2235-2241.

Chapter 8

Modeling Critical Forest Habitat in the Southern Coal Fields of West Virginia

Aaron E. Maxwell[1], Michael P. Strager[2], Charles B. Yuill[2], and J. Todd Petty[3]

[1]Natural Resource Analysis Center, West Virginia University, Morgantown, WV 26506, USA

[2]Division of Resource Management, West Virginia University, Morgantown, WV 26506, USA

[3]Division of Forestry and Natural Resources, West Virginia University, Morgantown, WV 26506, USA

ABSTRACT

Throughout the Central Appalachians of the United States resource extraction primarily from coal mining has contributed to the majority of the forest conversion to barren and reclaimed pasture and grass. The loss of forests in this Eco region is significantly impacting biodiversity at a regional scale. Since not all forest stands provide equal levels of ecological functions, it is critical to identify and map existing forested resources by the benefits that accrue from their unique spatial patterns, watershed drainage, and landscape positions. We utilized spatial analysis and remote sensing techniques to define critical forest characteristics. The characteristics were defined by applying a forest fragmentation model utilizing morphological image analysis, defining headwater catchments at a 1:24,000 scale, and deriving ecological land units (ELUs) from elevation data. Once critical forest values were calculated, it was possible to identify clusters of critical stands using spatial statistics. This spatially explicit method for modeling forest habitat could be implemented as a tool for assessing the impact of resource extraction and aid in the conservation of critical forest habitat throughout a landscape.

INTRODUCTION

Forests provide many benefits to society and the natural environment that include providing wildlife habitat [1], maintaining biodiversity [2], regulating climate and providing carbon storage [3, 4], nitrogen cycling [5,6], and altering and moderating hydrologic function, including evapotranspiration rates and surface runoff volumes [7]. Society land use needs and resource extraction often result in the loss of forested land cover, leading to an impact on biodiversity from habitat conversion [8–10]. In reference to forest and biodiversity conservation, there has been a shift in focus from rare or endangered species management to ecosystem and landscape scale management in which the health and function of the ecosystem as a whole is considered [11]. This focus requires that a diversity of ecological processes be considered, such as the health and structure of the forest [11]. The need is to focus on conservation of habitat richness and biodiversity at the landscape-level. Conservation efforts should be directed at spatially explicit, landscape-level attributes, such as minimizing structural contrast between adjacent landscape units [12]. The goal of this paper was to implement such a methodology in analyzing the conservation of biodiversity and forest habitat in the Southern Coal Fields of West Virginia.

We have expanded on previous work by suggesting and implementing a methodology that allows for the indexing of critical forest stands on a pixel-by-pixel basis relative to fragmentation patterns, watershed drainage, and landscape position. We implemented a methodology for extracting high-resolution, spatially explicit forest cover and disturbance data from National Agricultural Imagery Program (NAIP) orthophotography using object-based image classification and ancillary datasets. A fragmentation model utilizing morphological image analysis was then applied to the land cover data after Vogt et al. [13] to obtain high-resolution forest fragmentation data. These data were then related to watershed and landscape position in order to index forest pixels relative to importance of conservation for terrestrial habitat and biodiversity conservation within the region. An objective was to implement a spatial analysis method that utilized high-resolution data to describe the distribution and patterns of forest stands critical to biodiversity as a potential tool for evaluating and mitigating the effect of resource extraction in this unique and bio diverse landscape.

BACKGROUND

Not all forest stands are equal in terms of crucial habitat. Wickham et al. [14] documented the importance of core or interior forests in Southern Appalachia. Core and fragmented forests differ in terms of structure, composition, and ecological processes. Edge forests have higher rates of atmospheric decomposition [15] and higher proportions of exotic species [16]. In addition, it has been determined that the amount of forest cover is positively correlated with forest breeding bird populations [17]. Habitat edges have an influence on ecological function, including biogeochemical nutrient transport [18] and species interactions [19, 20]. Edges impact ecological mechanisms, patterns, and dynamics at variable spatial scales [21], including the landscape scale [22]. Fragmentation is of critical concern when modeling processes at the landscape scale [16].

Meyer et al. [23] documented the importance of headwater reaches. Such areas within a watershed contribute to overall biodiversity in a watershed, provide refuge from predators, competitors and alien species, and serve as links to adjacent watersheds. The unique conditions of an individual headwater reach contribute to the heterogeneity of the entire watershed and landscape and enhance biodiversity [24]. Appalachia supports 10% of the global salamander and freshwater mussel diversity [25, 26], and their diversity is high in headwater streams [27].

Straus Baugh and Core [28] documented the unique cove hardwoods or mixed mesophytic forests that support a wide range of species in Appalachia, such as a host of arachnid species [29, 30]. These communities are highly productive due to high moisture and fertile soil [31]. Cove hardwood forests are species rich [32, 33], and loss and fragmentation of these communities influence habitat suitability and potentially the persistence of the species on the landscape [31].

Weak land and Wood [34] and Wood ET. al. [35] documented the importance of upland, core forest for preservation and protection of the Cerulean Warbler, a species of concern that is threatened by habitat perturbations [34, 36, and 37]. As a result, we argue that forest fragmentation patterns, watershed position, and landscape position have an influence on how critical a forest stand is in terms of habitat conservation.

The Southern Coal Fields of West Virginia provides large expanses of temperate deciduous forest that are rare worldwide and provide for biodiversity [14]. For example, the eastern deciduous forest supports a wide variety of avian and large carnivore species; however, the extent and fragmentation of the forest have a direct impact on species abundance and composition [38]. In the temperate zone, the southern Appalachian region is among the highest in biodiversity and endemism with more than 2,000 documented vascular plant species and the highest freshwater biodiversity in North America [25, 39]. In the Southern Coal Fields of Appalachia, it has been suggested that increased seasonality of climate during the transition from glacial to interglacial conditions in the late Pleistocene and Holocene induced landscape instability that resulted in a heterogeneous mosaic of habitats [40]. This resulted in the rich and diverse biota observed today, and environmental changes can have a significant impact on the survival of species that have adapted to such specific conditions [40]. Southern Appalachian forests can be characterized as spatially extensive interior forests, and the extent of forest and its spatial patterns and distribution have been shown to have a direct influence on habitat suitability [41–43].

Land use and land disturbance impact forest extent, composition, and spatial patterns, which have consequences for biodiversity and other environmental services [44, 45]. Surface mining, and specifically mountain top removal with valley fills (MTR/VF), is currently the leading cause of land cover and land use change in the Southern Coal

Fields of West Virginia [46–49]. MTR/VF consists of clearing upland forests, removing top soil, and blasting away overburden rock material to extract coal seams. Overburden material is often placed in adjacent valleys, filling headwater steam segments. Land cover, landscape contours, and surface flow paths are altered [24]. Multiple watersheds in West Virginia have more than 10% of their surface area disturbed by surface mining [50]. The mining processes often results in a conversion of forested land cover to barren land cover. Later reclamation produces grasslands on the topographically altered mountaintops; however, productivity of the grasslands is often low due to poor soil conditions [51]. It has been estimated that all surface mining in Appalachia has resulted in a net loss of 420,000 ha of forest [48].

Wickham et al. [14] suggest that the amount of interior forest lost is greater than the amount of forest lost resulting from direct conversion of forest to surface mines. Forests loose interior character by the introduction of non-forest edges. Thus, the influence of surface mining extends into adjacent forest due to edge effects, and the spatially extensive nature of the forest is interrupted. Weak land and Wood [34] and Wood et al. [35] found that MTR/VF mining alters the spatial configuration of forested habitats, creating edge and area effects that negatively affect Cerulean Warbler abundance and occurrence on the landscape.

STUDY AREA

The study area for this project is a 2,280,423 ha area that encompasses the Southern Coal Fields of West Virginia (Figure 1). This area was defined relative to hydrologic unit code (HUC) 8 boundaries that intersect the MTR/VF region in West Virginia.

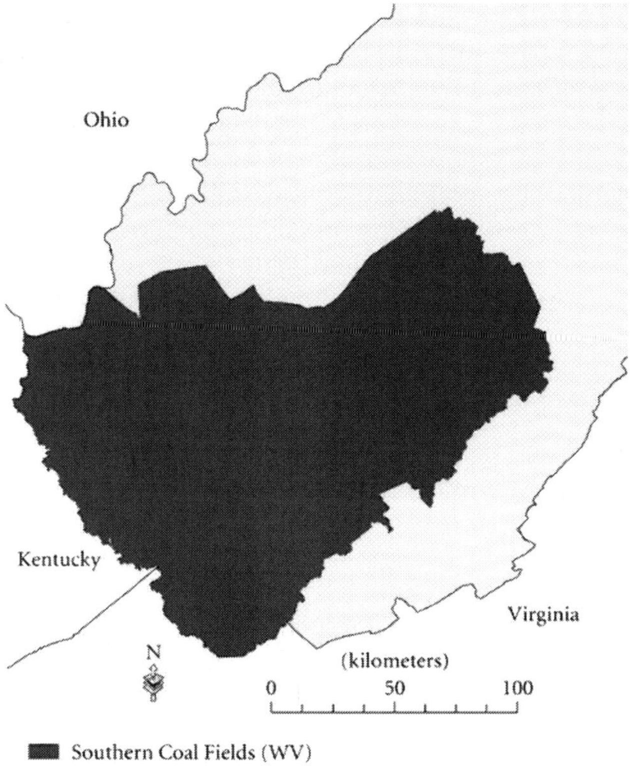

Figure 1: Study area. Southern Coal Fields of West Virginia.

The study area exists within the Appalachian Plateau physiogeographic province, a dissected, westward-tilted plateau dominated by Pennsylvanian bedrock. Pennsylvanian stratigraphy is characterized as cyclic sequences of sandstone, shale, clay, coal, and limestone [52]. The terrain is dissected by a dendritic stream network and shows fine texture with moderate-to-strong local relief. In comparison to the northern Appalachian Plateau, the Southern Coal Fields is generally more rugged due to resistant strata. Precipitation levels, temperature, and terrain allow for a variety of forest communities to flourish including oak-pine, oak-chestnut, cove hardwoods or mixed mesophytic forests, and floodplain communities [28]. Although surface mining has altered this landscape, it is still predominantly forested, in contrast to other regions of the eastern United States that have experienced development [46, 47].

METHODOLOGY

Indexing critical forest habitat incorporated several remote sensing image classification and spatial analysis processes. First, high-resolution land cover was extracted using object-based feature extraction and GIS decision rules. Object-based image classification allowed for the creation of high-resolution land cover/land use data from available aerial imagery. Second, a high-resolution forest fragmentation model was created for the region using a model developed by Vogt et al. [13]. Third, small catchment areas relative to 1:24,000 scale stream segments were extracted and headwater catchments were identified. And lastly, we incorporated digital elevation model (DEM) data at 1:4,800 scale to produce ecological land units (ELUs) for a thematic map representation of landforms. These products were then combined to describe critical forest habitat relative to fragmentation, watershed position, and landscape position.

Extraction of Land Cover/Land Use Data

Land cover/land use was produced utilizing an object-based feature extraction process along with decision rules. NAIP imagery was collected during leaf-on conditions in 2009 and made available in October 2009. The imagery has a 1 m nominal pixel spacing, a scale of 1:10,000, and is rectified to a horizontal accuracy of within ±5 meters of reference digital ortho quarter quads from the National Digital Ortho Program. This imagery was classified and results were merged to produce a continuous thematic dataset for the region.

Feature Analyst 4 by Visual Learning Systems (VLSs) within Erdas Imagine 9.3 was utilized to extract the following land cover types: barren/developed, grasslands, and forest. This software tool examines both the spectral and textural information within the imagery to produce thematic map data. As a result, unlike supervised classification techniques, feature extraction classifies an image using more than just the digital number (DN) or spectral information contained by each pixel. Spatial context information such as spatial association, size, shape, texture, pattern, and shadow are considered [53]. Studies have shown that feature extraction or objected-based algorithms are more effective and accurate at extracting information from high-resolution imagery than traditional image classification methods, such as supervised classification, because additional image characteristics are considered such as spatial context [54–57]. Feature extraction takes into account the spatial context that is available in high-resolution imagery, such as NAIP orthophotography. Due to the high spatial resolution and reduced spectral resolution of NAIP orthophotography in comparison to satellite imagery traditionally used to obtain thematic map data, we used feature extraction as the classification approach in order to make use of the spatial context and textural information contained in the imagery.

User-assisted feature extraction requires user-defined knowledge, as training data, to recognize and classify target features in an image [58]. As a result, we collected training samples as polygons by manually interpreting the imagery. For example, for a single dataset, we collected 101 examples of barren/developed land cover, 165 examples of grasslands land cover, and 147 examples of forested land cover. Our goal was to provide training data that accurately described the spectral and textural signature and variability of the land cover of interest.

An error assessment was performed by comparing the thematic map results to manual interpretation of the imagery at randomly selected locations. This assessment predicted 88% producer's accuracy and 99% user's accuracy for the forest class. As a result, this method allows for an accurate differentiation of forested land cover utilizing NAIP imagery to provide a recent and high spatial resolution representation of land cover and forest extent. It should be noted that datasets can be updated as new imagery is collected; for example, NAIP imagery has been collected in West Virginia in 2007, 2009, and 2011, which is a true benefit of utilizing NAIP. It is publically available, high-resolution data and is commonly flown every two years.

To further differentiate the land cover, we incorporated ancillary GIS data. Surface mine permit data from the West Virginia Department of Environmental Protection (WVDEP) were utilized to extract barren and grassland land cover that resulted from surface mining. For example, if barren land cover existed within a permit, it was determined to be a result of mining. This method allowed us to obtain an estimate of the area that is currently permitted but is still forested, and this classification represents areas on the landscape that have a higher potential for future disturbance relative to areas outside of surface mine permits.

This feature extraction and decision rule method allowed for the development of a high-resolution land cover dataset for the region relative to leaf-on 2009 conditions. Figure 2 describes the resulting land cover.

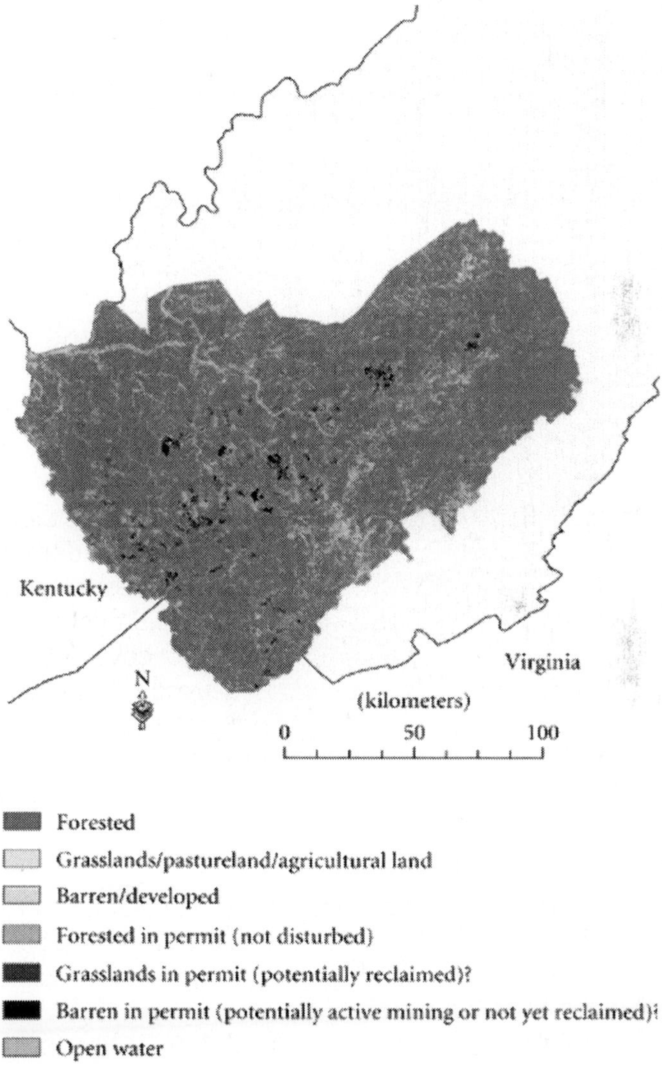

Figure 2: Land cover/land use data from object-based image classification of NAIP orthophotography.

Forest Fragmentation Model

The land cover data were reclassified as a binary forest/non-forest raster. The data were resampled using bilinear interpolation to a 9 m cell size to match the topographic data. Morphological image processing was applied to extract forest fragmentation data. The methodology is described in detail by Vogt et al. [13]. Morphological image processing uses mathematical morphology to analyze shape and form of objects [13]. Forest pixels are classified as patch, edge, perforated, or core using this model. This method has been shown to provide a more accurate representation of fragmentation at the single pixel or landscape level when compared to image convolution [13]. Figure 3 shows the fragmentation model.

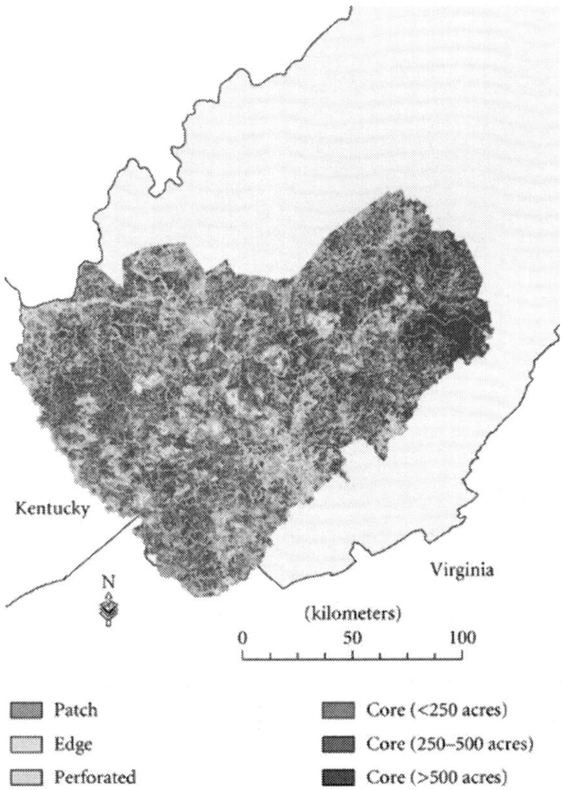

Figure 3: Fragmentation model after Vogt et al. [13].

Defining Watershed Position

Headwater watersheds were produced at a 1:24,000 scale as described by Stranger et al. [59]. The high-resolution National Hydrologic Dataset (NHD) stream segments were extracted and catchments delineated for all stream segments using DEM data. Photogram metrical elevation data from 1:4,800 scale were interpolated to a 3 m and then 9 m scale. Stream segment pour points were compared to a flow direction grid to obtain the catchment areas.

The high-resolution NHD data have tabulated flow networks and connectivity information for each segment, to delineate headwater reaches. These are reaches that do not collect drainage from upstream. As a result, the headwater catchments in this project are 1:24,000 scale catchments derived by comparing high-resolution NHD data to DEMs and were defined as catchments collecting no upstream drainage.

Modeling Landscape Position

The 9 m DEM data for the region, resampled from 3 m, was used to classify the terrain into the following ecological land units (ELUs): cliff, steep slope, slope crest, upper slope, flat summit, side slope, cove, dry flat, moist flat, wet flat, and slope bottom. The methodology described by Anderson et al. [60] provides a means to utilize DEM data to classify the landscape into different units of ecological significance. DEM derivatives were created including slope in degrees, a hierologically filled DEM, flow direction, and flow accumulation within GIS. The slope and flow accumulation grids were utilized to calculate a moisture index. The moisture index is a relative measure of the moisture of a specific cell, and it assumes that the moisture level is a function of how much water flows into the cell, predicted by flow accumulation, and how fast the water can flow out, described by slope [61–66].

Landscape position was calculated following Fels and Zobel [67], by dividing the landscape into the following categories: ridge, wide ridge, and slope/flat, slope/cove. The approach is based on a local neighborhood analysis in which a cell is compared to its neighbors. This method is based on distance-weighted assessment of elevation values in which a cell is compared to the mean elevation of all values within search windows of different sizes and extents. Landscape position was combined with slope and moisture index data to derive the ELU classes.

Once the ELU data were created, certain classes were selected as critical for our model. This included upland areas (slope crest and flat summit) and coves. This provided two mutually exclusive positions upon the landscape that were considered critical for biodiversity.

Critical Forest Model

The critical forest model was created by combining the core forest, forest in headwater catchments, cove forests, and upper slope forests. These layers were combined as binary raster layers to produce an index grid. Possible values ranged from 0 (not critical) to 3 (most critical). As the model was defined, cove and upper slope forests could not overlap, so the maximum index possible was 3. As a result, the final result of the analysis was an index grid that combines forest fragmentation, watershed position, and landscape position to describe forest critical for habitat and biodiversity at a 9 m pixel scale. The final output is shown in Figure 4. Figure 5 describes the overall process for the creation of the critical forest model.

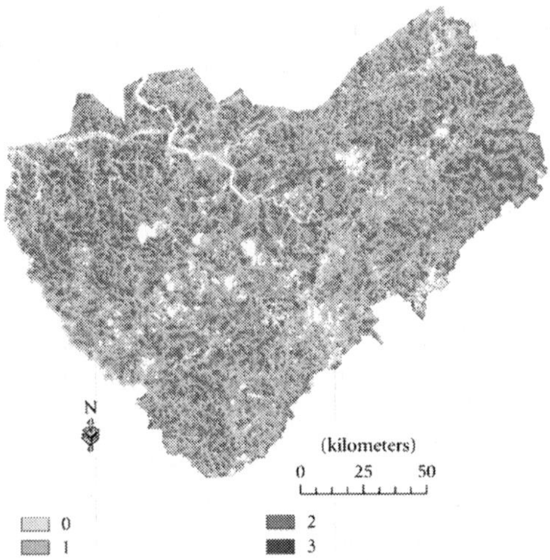

Figure 4: Critical forest model for Southern Coal Fields of West Virginia.

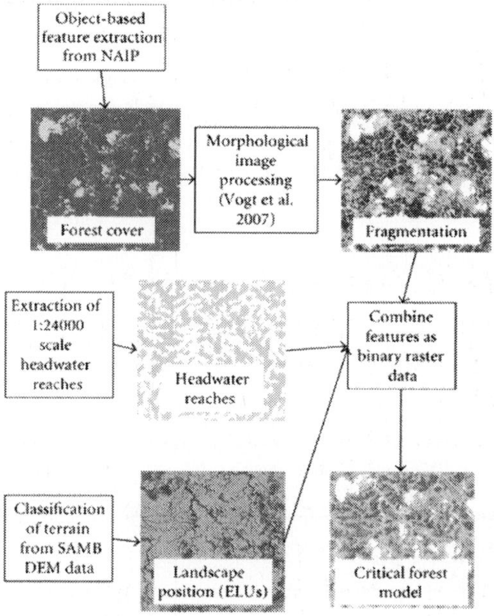

Figure 5: Flow chart for critical forest model creation.

Once the critical forest model was produced, we summarized the data relative to HUC (Hydrologic Unit Code) 8 and 11 catchments. We also assessed the data for spatial clustering using cluster and outlier analysis (Anselin Local Morans I) in GIS. Areas of surface mining, both active mining and reclamation, were converted to forested land cover, and the analysis was repeated using the potential premining land cover to analyze the distribution of premining critical forest habitat.

RESULTS

Land Cover/Land Use and Forest Fragmentation

The land cover/use for the entire study area extent is summarized in Table 1. Although this landscape has been heavily surface mined (contributing 3% of the land cover/land use), forest is still the dominant cover type (86.1% of the land cover); however, 3.2% of the landscape has been permitted for mining, but is currently not disturbed (forested). As a result, mine expansion is likely to induce further land cover change and loss of forest over large extents.

Table 1: Land cover/land use data for study area

Land cover/land use	Hectares	Percentage
Forested	1,892,774	82.9%
Grasslands/pastureland/agricultural land	172,316	7.5%
Barren/developed	54,770	2.4%
Forested in permit (not disturbed)	72,505	3.2%
Grasslands in permit (potentially reclaimed)	44,314	1.9%
Barren in permit (potentially active mining or not yet reclaimed)	25,621	1.1%
Open water	20,318	0.9%

The landscape disturbance is not evenly distributed throughout the region. Table 2 describes the land cover distribution for the Coal River HUC 8 watershed. According to our land cover calculations, 8.8% of the watershed is disturbed by active mining or reclamation while an additional 8.7% has been permitted but has not yet been disturbed or is still forested. The dominant land cover type in this watershed is forest; however, landscape disturbance due to surface mining is a dominant land use feature.

Table 2: Land cover/land use data for the Coal River Watershed

Land cover/land use	Hectares	Percentage of watershed
Forested	168,420	73.0%
Grasslands/pastureland/agricultural land	1,6420	7.1%
Barren/developed	4,140	1.8%
Forested in permit (not disturbed)	20,070	8.7%
Grasslands in permit (potentially reclaimed)	12,760	5.5%

Barren in permit (potentially active mining or not yet reclaimed)	7,640	3.3%
Open water	1,310	0.6%

Of the forested land cover within the region, 51% was predicted as core forest using the Vogt et al. [13] model, 38.1% was classified as perforated, 9.2% as edge, and 1.2% as patch. Of the core forest, the majority of the pixels were within large core extents, greater than 500 acres. Forested areas currently in permits were predicted as dominantly being edge and perforated forest explained possibly by the adjacency to active mine sites and reclamation or small forest stands surrounded by reclamation.

Critical Forest Model

Table 3 shows the distribution of critical forest index values throughout the region and only within the permitted forested areas. Based on visual interpretation of the results, the most critical areas as defined by the model are either cove or upper slope core forests within headwater reaches. As can be seen in Table 1, it was very common for a pixel to satisfy at least one of the criteria.

Table 3: Critical forest index distributions

Index	Study area		Forest in Permit	
	Hectares	Percentage	Hectares	Percentage
0	351,924	17.7%	14,406	21.5%
1	819,975	41.3%	35,768	53.3%
2	670,289	33.8%	15,305	22.8%
3	143,474	7.2%	1,596	2.4%

The data were summarized relative to HUC 8 and HUC 11 watershed that intersect the MTR/VF region and exist entirely within West Virginia. This data suggests that The Elk River HUC 8 watershed has the highest percentage of critical forest value pixels (values 1 through 3) within the MTR/VF region of West Virginia. At the HUC 11 scale, we found that local, spatial clusters of watersheds exist when the percentage of critical forest is considered.

Perhaps a more useful representation of spatial clustering for land management and regulation purposes would be the stand scale throughout the landscape as opposed to large catchment areas. We smoothed the raster critical forest model using majority filter operations within GIS then converted contiguous areas of a single critical forest value to polygon features, and a cluster and outlier analysis of these areas was performed relative to area of the polygons and critical forest value. This showed that clusters of critical forest areas exist across the landscape, especially when value 3 areas are of interest. Clusters of critical forest stands could represent areas that should be managed or protected in order to preserve biodiversity throughout the landscape and region.

Table 4 describes the distribution of critical forest values in areas of mine disturbance and it offers an estimate of the critical nature of mine disturbed areas based on the assumption that these areas were forested on the premining landscape, which we argue is a valid assumption based on comparisons to premining satellite data and the landscape position of the mines. Figure 6 offers a comparison of pre- and post-mining conditions. This data suggests that critical forest areas overlap with areas of resource extraction. It has been documented that mining impacts headwater catchments and upland terrains, which are critical for biodiversity [14, 24, and 35]. This data suggests that there is a need to delineate stands and clusters of critical forests habitats throughout the region because they are being deforested and impacted by resource extraction.

Table 4: Critical forest index distribution in mined areas

Value	Hectares	Percentage
0	61,949.7	9.3%
1	222,677.1	33.4%
2	296,801.1	44.5%
3	84,903.3	12.7%

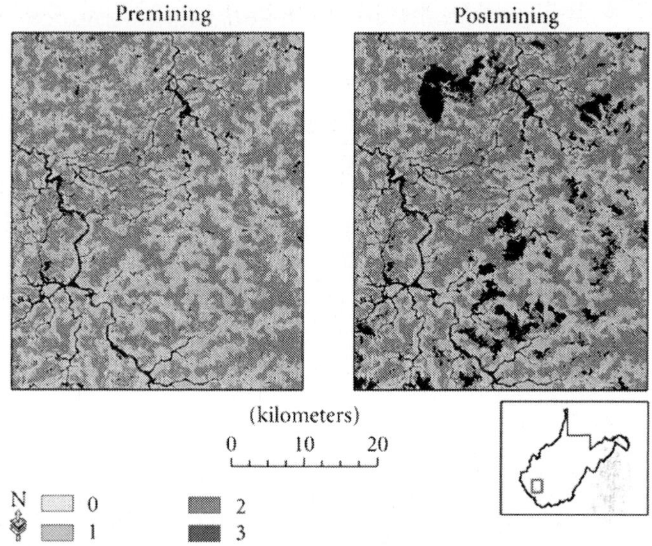

Figure 6: Pre- and post-mining critical forest habitat.

DISCUSSION

Our critical forest modeling is based on that all forest stands are not equal in terms of habitat suitability and supporting biodiversity. By taking into account high-resolution forest fragmentation patterns, watershed position, and landscape position, forest pixels were indexed relative to these factors that have an impact on habitat. Continued surface mining in the region will likely continue to alter forests, habitats, and the ecosystem and landscape as a whole, and this practice has been shown to produce both aquatic and terrestrial habitat degradation [24, 50]. Perhaps considering not just the extent of forest to be disturbed, but the ecological function of the forest, is necessary when continued mining and forest disturbance are proposed. Perhaps certain forests, such as core, headwater, cove forests, should be protected over other forested areas that are not as ecologically valuable. As resource extraction continues, perhaps regulators should take into account such abiotic factors of the landscape and adjust practices to minimize disturbance of critical areas. Perhaps spatial clusters of critical forest habitat should be considered during the permitting process.

We argue that a spatially explicit and high-resolution means of classifying or indexing critical forest stands is valuable as a regulatory tool for assessing the impact of continued mining and resource extraction in this region. McDermid et al. [68] suggest that remote sensing and GIS data that consider multiple attributes for modeling wildlife habitat are valuable, especially for large study areas. They also highlighted the use of land cover derived from object-based image classification, as was performed here. Their study focused on explaining patterns of grizzly bears habitat use within a study area in Alberta, Canada, and they found that incorporating data beyond categorical land cover was superior for their modeling purposes. Currently, there is a lack of guidelines for the selection and analysis of spatial data for exploring environmental and ecological questions [68]. We suggest that considering forest fragmentation (a derivative of land cover) along with catchment and terrain-derived variables provided a better understanding of critical habitat in the Southern Coal Fields of West Virginia than could be obtained utilizing land cover alone.

Optimal spatial resolution of thematic map data for a region of this extent and for this mapping purpose is debatable. As an example, Landsat MSS data, offering a 60 m pixel, has been shown to be inadequate for urban mapping; also, as a general rule a pixel should be at least one-half the size of the shortest dimension of the smallest feature that is being mapped [69]. Our aim was to describe the forest cover at a stand level and to map disturbance features with small spatial extents that interrupt the canopy. As a result, a finer resolution dataset was required than what Landsat Multispectral Scanner (MSS) or Thematic Mapper (TM) data could offer; as a result, NAIP imagery and object-based classifiers were utilized to obtain the base thematic map data from which the forest cover and forest fragmentation were derived. Now that object-based image classification tools are available and high-resolution land cover and forest cover can be extracted from high-resolution and low spectral resolution imagery, forest cover data at the stand scale may become more widely available. We note that there is value in including such data in environmental modeling and assessment practices and combining that data with other datasets, such as topographic data, to model multiple attributes of the landscape.

Forest alteration resulting from mountaintop mining may not be characteristic of other human land use practices that result in deforestation, such as agriculture. Generally, forest cover change is most common in areas of low elevation and gentle terrain [70]. Mountaintop mining often induces land cover change in upland areas, headwater reaches, and rugged terrains [24, 35]. As a result, there is a need to understand the critical nature of these upland forests in order to assess the environmental impact of mountaintop mining, which produces unique landscape alteration.

Critical forest habitat modeling may also be useful in other landscapes impacted by resource extraction. For example, extraction of natural gas from the Marcellus Shale Formation throughout parts of New York, Pennsylvania, West Virginia, and Maryland in the eastern United States could contribute to habitat fragmentation and alteration of forest extents as access roads and well pads are established [71]. High-resolution land cover, fragmentation, watershed, and terrain data could be used to model the impact and predict critical habitats using similar methodology as proposed here.

CONCLUSIONS

We have presented a method for indexing of critical forest stands on a pixel-by-pixel basis relative to fragmentation patterns, watershed drainage, and landscape position. This methodology was implemented throughout the Southern Coal Fields of West Virginia and could be applied in other regions impacted by resource extraction. It relies on high-resolution spatially explicit forest cover from NAIP imagery using object-based feature extraction, creating a forest fragmentation model after Vogt et al. [13] utilizing morphological image analysis, defining headwater catchments at a 1:24,000 scale, and extracting ecological land units (ELUs) from DEM data. Once critical forest values are calculated, it is possible to identify spatial clusters of critical stands. This spatially explicit method for modeling forest habitat could be implemented as a tool for assessing the impact of resource extraction and aid in the conservation of critical habitat.

Future research should focus on incorporating additional landscape variables to model and differentiate forest stands. For example, canopy height data derived from LiDAR (light detection and ranging) could provide additional forest metrics for modeling. Additional pattern and connectivity analysis may provide for a better understanding of the network of critical forest stands throughout the region.

Our goal was to present a remote sensing and GIS method for indexing forest pixels throughout a disturbed landscape and was based on the assumption that all forests are not equal in terms of habitat. As resource extraction continues throughout the region, considering the nature of the forest to be disturbed may aid in mitigating environmental impacts of human land use.

REFERENCES

1. D. D. Chiras, J. P. Reganold, and O. S. Owen, Natural Resource Conservation, Prentice Hall, Upper Saddle River, NJ, USA, 8th edition, 2002.
2. T. E. Lovejoy, R. O. Bierregaard, A. B. Rylands, et al., "Edge and other effects of isolation on Amazonian forest fragmentation," in Conservation Biology: The Science of Scarcity and Diversity, M. Soule, Ed., pp. 257–285, Sinauer Associates, Sunderland, Mass, USA, 1986.
3. R. R. Nemani and S. W. Running, "Satellite monitoring of global land cover changes and their impact on climate," Climatic Change, vol. 31, no. 2–4, pp. 395–413, 1995.
4. R. A. Houghton, J. L. Hackler, and K. T. Lawrence, "The U.S. carbon budget: contributions from land-use change," Science, vol. 285, no. 5427, pp. 574–578, 1999.
5. P. M. Vitousek, J. D. Aber, R. W. Howarth et al., "Human alteration of the global nitrogen cycle: sources and consequences," Ecological Applications, vol. 7, no. 3, pp. 737–750, 1997.
6. S. R. Carpenter, N. F. Caraco, D. L. Correll, R. W. Howarth, A. N. Sharpley, and V. H. Smith, "Nonpoint pollution of surface waters with phosphorus and nitrogen," Ecological Applications, vol. 8, no. 3, pp. 559–568, 1998.
7. R. E. Dickinson, "Global change and terrestrial hydrology—a review," Tellus, vol. 43, no. 4, pp. 176–181, 1991.
8. J. Chow, R. J. Kopp, and P. R. Portney, "Energy resources and global development," Science, vol. 302, no. 5650, pp. 1528–1531, 2003.
9. A. P. Dobson, A. D. Bradshaw, and A. J. M. Baker, "Hopes for the future: restoration ecology and conservation biology," Science, vol. 277, no. 5325, pp. 515–522, 1997.

10. W. E. Westman, "Managing for biodiversity: unresolved science and policy questions," BioScience, vol. 40, no. 1, pp. 26–33, 1990.
11. K. A. Poiani, B. D. Richter, M. G. Anderson, and H. E. Richter, "Biodiversity conservation at multiple scales: functional sites, landscapes, and networks," BioScience, vol. 50, no. 2, pp. 133–146, 2000.
12. J. F. Franklin, "Preserving biodiversity: species, ecosystems, or landscape?" Ecological Applications, vol. 3, no. 2, pp. 202–205, 1993.
13. P. Vogt, K. H. Riitters, C. Estreguil, J. Kozak, T. G. Wade, and J. D. Wickham, "Mapping spatial patterns with morphological image processing," Landscape Ecology, vol. 22, no. 2, pp. 171–177, 2007.
14. J. D. Wickham, K. H. Riitters, T. G. Wade, M. Coan, and C. Homer, "The effect of Appalachian mountaintop mining on interior forest," Landscape Ecology, vol. 22, no. 2, pp. 179–187, 2007.
15. K. C. Weathers, M. L. Cadenasso, and S. T. A. Pickett, "Forest edges as nutrient and pollutant concentrators: potential synergisms between fragmentation, forest canopies, and the atmosphere,"Conservation Biology, vol. 15, no. 6, pp. 1506–1514, 2001.
16. K. A. Harper, S. E. MacDonald, P. J. Burton et al., "Edge influence on forest structure and composition in fragmented landscapes," Conservation Biology, vol. 19, no. 3, pp. 768–782, 2005.
17. M. K. Trzcinski, L. Fahrig, and G. Merriam, "Independent effects of forest cover and fragmentation on the distribution of forest breeding birds," Ecological Applications, vol. 9, no. 2, pp. 586–593, 1999.
18. J. F. Kitchell, R. V. O'Neil, D. Webb, et al., "Consumer regulation of nutrient cycling," BioScience, vol. 29, no. 1, pp. 28–34, 1979.
19. P. Kareiva and G. Odell, "Swarms of predators exhibit "preytaxis" if individual predators use area-restricted search," American Naturalist, vol. 130, no. 2, pp. 233–270, 1987.
20. J. Roland and P. D. Taylor, "Insect parasitoid species respond to forest structure at different spatial scales," Nature, vol. 386, no. 6626, pp. 710–713, 1997.

21. T. B. Smith, R. K. Wayne, D. J. Girman, and M. W. Bruford, "A role for ecotones in generating rainforest biodiversity," Science, vol. 276, no. 5320, pp. 1855–1857, 1997.
22. T. M. Donovan, P. W. Jones, E. M. Annand, and F. R. Thompson III, "Variation in local-scale edge effects: mechanisms and landscape context," Ecology, vol. 78, no. 7, pp. 2064–2075, 1997.
23. J. L. Meyer, D. L. Strayer, J. B. Wallace, S. L. Eggert, G. S. Helfman, and N. E. Leonard, "The contribution of headwater streams to biodiversity in river networks," Journal of the American Water Resources Association, vol. 43, no. 1, pp. 86–103, 2007.
24. E. S. Bernhardt and M. A. Palmer, "The environmental costs of mountaintop mining valley fill operations for aquatic ecosystems of the Central Appalachians," Annals of the New York Academy of Sciences, vol. 1223, no. 1, pp. 39–57, 2011.
25. L. L. Master, S. R. Flack, and B. A. Stein, Rivers of Life: Critical Watersheds for Protecting Freshwater Biodiversity, The Nature Conservancy, Arlington, Va, USA, 1998.
26. N. B. Green and T. K. Pauley, Amphibians and Reptiles in West Virginia, University of Pittsburgh Press, Pittsburgh, Pa, USA, 1987.
27. A. Clarke, R. Mac Nally, N. Bond, and P. S. Lake, "Macroinvertebrate diversity in headwater streams: a review," Freshwater Biology, vol. 53, no. 9, pp. 1707–1721, 2008.
28. P. D. Strausbaugh and E. L. Core, Flora of West Virginia, Seneca Books, Morgantown, WVa, USA, 1977.
29. J. A. Coddington, L. H. Young, and F. A. Coyle, "Estimating spider species richness in a Southern Appalachian cove hardwood forest," The Journal of Arachnology, vol. 24, no. 2, pp. 111–128, 1996.
30. F. A. Coyle, "Effects of clearcutting on the spider community of a Southern Appalachian forest," The Journal of Arachnology, vol. 9, pp. 285–298, 1981.
31. M. G. Turner, S. M. Pearson, P. Bolstad, and D. N. Wear, "Effects of land-cover change on spatial pattern of forest communities in the Southern Appalachian Mountains (USA)," Landscape Ecology, vol. 18, no. 5, pp. 449–464, 2003.
32. S. M. Pearson, A. B. Smith, and M. G. Turner, "Forest patch size, land use, and mesic forest herbs in the French Broad River Basin, North Carolina," Castanea, vol. 63, no. 3, pp. 382–395, 1998.

33. C. E. Mitchell, M. G. Turner, and S. M. Pearson, "Effects of historical land use and forest patch size on myrmechocores and ant communities in the Southern Appalachian Highlands (USA)," Ecological Applications, vol. 12, pp. 1364–1377, 2002.
34. C. A. Weakland and P. B. Wood, "Cerulean warbler (Dendroica cerulea) microhabitat and landscape-level habitat characteristics in Southern West Virginia," The Auk, vol. 122, no. 2, pp. 497–508, 2005.
35. P. B. Wood, S. B. Bosworth, and R. Dettmers, "Cerulean warbler abundance and occurrence relative to large-scale edge and habitat characteristics," The Condor, vol. 108, no. 1, pp. 154–165, 2006.
36. P. B. Hamel, D. K. Dawson, and P. D. Keyser, "How we can learn more about the cerulean warbler (Dendroica cerulea)," The Auk, vol. 121, no. 1, pp. 7–14, 2004.
37. J. Jones, R. D. DeBruyn, J. J. Barg, and R. J. Robertson, "Assessing the effects of natural disturbance on a neotropical migrant songbird," Ecology, vol. 82, no. 9, pp. 2628–2635, 2001.
38. P. A. Keddy and C. G. Drummond, "Ecological properties for the evaluation, management, and restoration of temperate deciduous forest ecosystems," Ecological Applications, vol. 6, no. 3, pp. 748–762, 1996.
39. B. A. Stein, L. S. Kutner, and J. S. Adams, Precious Heritage: The Status of Biodiversity in the United States, Oxford University Press, New York, NY, USA, 2000.
40. P. A. Delcourt and H. R. Delcourt, "Paleoecological insights on conservation of biodiversity: a focus on species, ecosystems, and landscapes," Ecological Applications, vol. 8, no. 4, pp. 921–934, 1998.
41. SAMAB (Southern Appalachian Man and the Biosphere), "The Southern Appalachian assessment terrestrial," Tech. Rep. Report 5 of 5, US Department of Agriculture, Forest Service, Atlanta, Ga, USA, 2008, http://www.samab.org/saa/saa_reports.html.
42. S. K. Robinson, F. R. Thompson III, T. M. Donovan, D. R. Whitehead, and J. Faaborg, "Regional forest fragmentation and the nesting success of migratory birds," Science, vol. 267, no. 5206, pp. 1987–1990, 1995.

43. L. Fahrig, "Effect of habitat fragmentation on the extinction threshold: a synthesis," Ecological Applications, vol. 12, no. 2, pp. 346–353, 2002.
44. D. J. Mladenoff, M. A. White, J. Pastor, and T. R. Crow, "Comparing spatial pattern in unaltered old-growth and disturbed forest landscapes," Ecological Applications, vol. 3, no. 2, pp. 294–306, 1993.
45. K. H. Riitters, J. D. Wickham, R. V. O'Neill et al., "Fragmentation of continental United States forests,"Ecosystems, vol. 5, no. 8, pp. 815–822, 2002.
46. R. L. Hooke, "Spatial distribution of human geomorphic activity in the United States: comparison with rivers," Earth Surface Processes and Landforms, vol. 24, no. 8, pp. 687–692, 1999.
47. K. L. Saylor, "Land Cover Trends Project: Central Appalachians," U.S. Department of the Interior, U.S. Geological Survey, Washington, DC, USA, 2008,http://landcovertrends.usgs.gov/east/eco69Report.html.
48. M. A. Drummond and T. R. Loveland, "Land-use pressure and a transition to forest-cover loss in the Eastern United States," BioScience, vol. 60, no. 4, pp. 286–298, 2010.
49. P. A. Townsend, D. P. Helmers, C. C. Kingdon, B. E. McNeil, K. M. de Beurs, and K. N. Eshleman, "Changes in the extent of surface mining and reclamation in the Central Appalachians detected using a 1976–2006 Landsat time series," Remote Sensing of Environment, vol. 113, no. 1, pp. 62–72, 2009.
50. M. A. Palmer, E. S. Bernhardt, W. H. Schlesinger et al., "Mountaintop mining consequences," Science, vol. 327, no. 5962, pp. 148–149, 2010.
51. J. A. Simmons, W. S. Currie, K. N. Eshleman et al., "Forest to reclaimed mine land use change leads to altered ecosystem structure and function," Ecological Applications, vol. 18, no. 1, pp. 104–118, 2008.
52. WVGES (West Virginia Geologic and Economic Survey), "Physiographic provinces of West Virginia," 2005, http://www.wvgs.wvnet.edu/www/maps/pprovinces.htm.
53. D. Opitz, "An automated change detection system for specific features," in Proceedings of the International ESRI User

Conference, Environmental Systems Research Institute, San Diego, Calif, USA, July 2003.

54. M. A. Friedl and C. E. Brodley, "Decision tree classification of land cover from remotely sensed data,"Remote Sensing of Environment, vol. 61, no. 3, pp. 399–409, 1997.

55. N. R. Harvey, J. Theiler, S. P. Brumby et al., "Comparison of GENIE and conventional supervised classifiers for multispectral image feature extraction," IEEE Transactions on Geoscience and Remote Sensing, vol. 40, no. 2, pp. 393–404, 2002.

56. G. Hong, Y. Zhang, and D. A. Lavigne, "Object-based change detection in high resolution image," inProceedings of the American Society for Photogrammetry and Remote Sensing Annual Conference (ASPRS '06), American Society for Photogrammetry and Remote Sensing, Reno, Nev, USA, May 2006.

57. J.V. Kaiser Jr., D. A. Stow, and L. Cao, "Evaluation of remote sensing techniques for mapping transborder trails," Photogrammetric Engineering and Remote Sensing, vol. 70, no. 12, pp. 1441–1447, 2004.

58. Visual Learning Systems, "User Manual, Feature Analyst 4.2 for ERDAS Imagine," Visual Learning Systems, Missoula, Mont, USA, p. 354, 2002,http://www.featureanalyst.com/feature_analyst/publications/manuals/FA_4.2_RELEASE/FA_4.2_Reference_arc_040908_RELEASE.pdf.

59. M. P. Strager, J. T. Petty, J. M. Strager, and J. Barker-Fulton, "A spatially explicit framework for quantifying downstream hydrologic conditions," Journal of Environmental Management, vol. 90, no. 5, pp. 1854–1861, 2009.

60. M. G. Anderson, M. D. Merrill, and F. B. Biasi, Connecticut River Watershed Analysis: Ecological Communities and Neo-Tropical Migratory Birds (Final Report), Eastern Conservation Science, the Nature Conservancy, 1998.

61. R. B. Grayson, I. D. Moore, and T. A. McMahon, "Physically based hydrologic modeling 1. A terrain-based model for investigative purposes," Water Resources Research, vol. 28, no. 10, pp. 2639–2658, 1992.

62. H. Mitasova, J. Hofierka, M. Zlocha, and L. R. Iverson, "Modelling topographic potential for erosion and deposition using GIS,"

International Journal of Geographical Information Systems, vol. 10, no. 5, pp. 629–641, 1996.

63. D. M. Moore, B. G. Lees, and S. M. Davey, "A new method for predicting vegetation distributions using decision tree analysis in a geographic information system," Environmental Management, vol. 15, no. 1, pp. 59–71, 1991.
64. M. Boer, G. Del Barrio, and J. Puigdefabregas, "Mapping soil depth classes in dry Mediterranean areas using terrain attributes derived from a digital elevation model," Geoderma, vol. 72, no. 1-2, pp. 99–118, 1996.
65. E. M. O'Loughlin, "Prediction of surface saturation zones in natural catchments by topographic analysis," Water Resources Research, vol. 22, no. 5, pp. 794–804, 1986.
66. A. J. Parker, "The topographic relative moisture index: an approach to soil- moisture assessment in mountain terrain," Physical Geography, vol. 3, no. 2, pp. 160–168, 1982.
67. J. Fels and R. Zobel, "Landscape position and classified land type mapping for the statewide DRASTIC mapping project," Tech. Rep. VEL.95.1, North Carolina State University, North Carolina Department of Environmental, Health, and Natural Resources, Division of Environmental Management, 1995.
68. G. J. McDermid, R. J. Hall, G. A. Sanchez-Azofeifa et al., "Remote sensing and forest inventory for wildlife habitat assessment," Forest Ecology and Management, vol. 257, no. 11, pp. 2262–2269, 2009.
69. J. R. Jensen, Introductory Digital Image Processing: A Remote Sensing Perspective, Pearson Prentice Hall, Upper Saddle River, NJ, USA, 3rd edition, 2005.
70. D. N. Wear and R. O. Flamm, "Public and private forest disturbance regimes in the Southern Appalachians," Natural Resource Modeling, vol. 7, no. 4, pp. 379–397, 1993.
71. D. J. Soeder, "The marcellus shale: resources and reservations," EOS Transactions American Geophysical Union, vol. 91, no. 32, pp. 277–288, 2010.

Citations

CHAPTER 1

L. Dupin, C. Nkono, C. Burlet, F. Muhashi and Y. Vanbrabant, "Land Cover Fragmentation Using Multi-Temporal Remote Sensing on Major Mine Sites in Southern Katanga (Democratic Republic of Congo)," Advances in Remote Sensing, Vol. 2 No. 2, 2013, pp. 127-139. doi: 10.4236/ars.2013.22017.

CHAPTER 2

Effah K Antwi, John Boakye-Danquah, Stephen B Asabere, Kazuhiko Takeuchi, and Gerhard Wiegleb, Land cover transformation in two post-mining landscapes subjected to different ages of reclamation since dumping of spoils, doi:10.1186/2193-1801-3-702.

CHAPTER 3

Isaac O. Osunmakinde, "Towards Safety from Toxic Gases in Underground Mines Using Wireless Sensor Networks and Ambient Intelligence," International Journal of Distributed Sensor Networks, vol. 2013, Article ID 159273, 15 pages, 2013. doi:10.1155/2013/159273.

CHAPTER 4

F. Feizi and E. Mansouri, "Introducing the Iron Potential Zones Using Remote Sensing Studies in South of Qom Province, Iran," *Open Journal of Geology*, Vol. 3 No. 4, 2013, pp. 278-286. doi: 10.4236/ojg.2013.34032.

CHAPTER 5

Cheng Bo, Cheng Xin, Zhai Zhongyi, Zhang Chengwen, and Chen Junliang, "Web of Things-Based Remote Monitoring System for Coal Mine Safety Using Wireless Sensor Network," International Journal of Distributed Sensor Networks, vol. 2014, Article ID 323127, 14 pages, 2014. doi:10.1155/2014/323127.

CHAPTER 6

Tingting Liu, Liangpei Zhang, Pingxiang Li, and Hui Lin, Remotely sensed image retrieval based on region-level semantic mining, doi:10.1186/1687-5281-2012-4.

CHAPTER 7

I-Chen Wu, Siang-Rou Lu, and Bin-Chen Hsiung, A BIM-Based Monitoring System for Urban Deep Excavation Projects, doi:10.1186/s40327-014-0015-x.

CHAPTER 8

Aaron E. Maxwell, Michael P. Strager, Charles B. Yuill, and J. Todd Petty, "Modeling Critical Forest Habitat in the Southern Coal Fields of West Virginia," International Journal of Ecology, vol. 2012, Article ID 182683, 10 pages, 2012. doi:10.1155/2012/182683.

Index

V

W